Mothers and the Mexican Antinuclear Power Movement

Society, Environment, and Place

Series Editors: Andrew Kirby and Janice Monk

Mothers and the Mexican Antinuclear Power Movement

Velma García-Gorena

The University of Arizona Press   Tucson

The University of Arizona Press
© 1999 The Arizona Board of Regents
First printing
All rights reserved
∞ This book is printed on acid-free, archival-quality paper.
Manufactured in the United States of America
04 03 02 01 00 99   6 5 4 3 2 1

García-Gorena, Velma.
Mothers and the Mexican antinuclear power movement/Velma García-Gorena.
p. cm. – (Society, environment, and place)
Includes bibliographical references and index.
ISBN 0-8165-1874-2 (acid-free paper)
ISBN 0-8165-1875-0 (pbk.: acid-free paper)
1. Antinuclear movement–Mexico–Laguna Verde (Veracruz)
2. Nuclear power plants–Mexico–Laguna Verde (Veracruz)
3. Social movements–Mexico–Laguna Verde (Veracruz)
4. Women social reformers–Mexico–Laguna Verde (Veracruz)
I. Title. II. Series.
HD9698.M63 V474 1999   98-25504
327.1´747´097262—ddc21   CIP

British Library Cataloguing-in-Publication Data
A catalogue record for this book is available from the British Library.

Publication of this book is made possible in part by the proceeds of a
permanent endowment created with the assistance of a Challenge
Grant from the National Endowment for the Humanities,
a federal agency.

For Ruth, David, and the rest of my family—on both sides of the border

Contents

Preface

I was in Xalapa, Veracruz, Mexico, in July 1988 studying a peasant movement when friends commented about their participation in a dramatic, three-day highway blockade to protest the construction of a nuclear power plant—Laguna Verde—located a few miles from the port of Veracruz on the Gulf of Mexico. One of my friends, Sara González, told me that she had decided to give up volunteering for the Red Cross because of her work as one of the founding members of a brand new antinuclear group, the Madres Veracruzanas, which had been modeled after the Madres de la Plaza de Mayo in Argentina. A middle-class housewife and grandmother, she showed me boxes full of newspaper clippings and other materials documenting the short history of the Mexican antinuclear movement. She explained that many of the residents of Veracruz were fully aware of what had happened at Chernobyl and that local mothers had decided to try to protect their children from the consequences of a similar accident possible at Laguna Verde.

At the time, I was not aware that Mexico *had* a nuclear energy program, so I immersed myself in the materials she had shown me. I also attended a meeting of the Madres Veracruzanas and was impressed by the members' knowledge of nuclear energy and the Mexican political system. They had already participated in numerous protests, had met with several important government officials, and were at this point demanding a meeting with President Miguel de la Madrid. Shortly after this meeting, I gave up my other project and began studying this movement against the Laguna Verde plant.

I have now spent years engaged in research on the Mexican antinuclear movement. A Ford Foundation Postdoctoral Fellowship for Minorities allowed me to spend the academic year 1989–90 in Mexico studying various antinuclear organizations. I also returned to Mexico every summer between 1990 and 1995. A Picker Fellowship at Smith College allowed me to complete the manuscript. I spent a great deal of time observing movement participants—during the weekly Saturday protests in Xalapa, at discussions in the Chamber of Deputies in Mexico City, at marches in the village of Palma Sola, and at various other events and places. I also conducted numerous interviews with movement participants. I also relied on newspaper accounts of the movement. Although the Mexican government imposed a heavy censorship on television coverage of the movement, several local and national newspapers provided excellent coverage of movement activities and of the government's reaction to this mobilization.

I am indebted to many people for helping me with my research and writing. In Xalapa, Sara González, Rebeka Dyer, Patricia Ortega, Mirna Benítez, Mercedes Solé, and Pedro Lizárraga provided extensive information about the movement and waited patiently for me to finish the manuscript. In addition, Sara González spent several months in newspaper archives verifying footnotes, and Rebeka Dyer read and provided feedback on the entire manuscript. Rafael Ruiz provided invaluable help by directing me to a CD-ROM database of newspaper articles. I also wish to thank Sara Ladrón de Guevara for her support and for directing me to very important sources of information. My colleagues at Smith College—Susan Bourque, Martha Ackelsberg, Donna Divine, Nancy Whittier, and Gary Lehring—also provided invaluable advice on the final steps of manuscript preparation. Anna Fedyshyn and Martha Frye also provided excellent editorial help. In addition, my fellow Mexicanists, Rebecca Horn and Marjorie Becker, provided unwavering support, and I am very grateful. I also wish to thank the editors at the University of Arizona Press—Annie Barva, Alan Schroder, and Christine Szuter—for their help and guidance. Finally, two anonymous reviewers provided extremely helpful and thoughtful comments.

Abbreviations

AE	Alianza Ecologista
ATA	Asociación de Tecnología Apropiada (Association for Appropriate Technology)
CIR	Centro de Ingeniería de Reactores (Center for Reactor Engineering)
CFE	Comisión Federal de Electricidad (Federal Electricity Commission)
CNSNS	Comisión Nacional de Seguridad Nuclear y Salvaguardas (National Commission on Nuclear Safety)
CNOP	Confederación Nacional de Organizaciones Populares (National Confederation of Popular Organizations)
COCEI	Coalición de Obreros, Campesinos, y Estudiantes del Istmo (Coalition of Workers, Peasants, and Students of the Isthmus)
CONCLAVE	Coordinadora Nacional Contra Laguna Verde (National Coordinating Committee against Laguna Verde)
FDM	Frente Democrático Nacional (National Democratic Front)
MAEV	Movimiento Antinuclear del Estado de Veracruz (Antinuclear Movement of the State of Veracruz)
MAMA	Mujeres Activas Mexicanas Antinucleares (Active Antinuclear Mexican Women)
MEM	Movimiento Ecologista Mexicano (Mexican Ecology Movement)

NSM	New Social Movement
PAN	Partido de Acción Nacional (National Action Party)
PARM	Partido Auténtico de la Revolución Mexicana (Authentic Party of the Mexican Revolution)
PERE	Plan de Emergencia Radiológica Externo (Emergency Evacuation Plan)
PFCRN	Partido Frente Cardenista de Reconstrucción Nacional (Cardenista Front for National Reconstruction)
PPS	Partido Popular Socialista (Popular Socialist Party)
PRD	Partido de la Revolución Democrática (Party of the Democratic Revolution)
PRI	Partido Revolucionario Institucional (Institutional Revolutionary Party)
SEDUE	Secretaría de Desarrollo Urbano y Ecología (Ministry for Urban Development and the Environment)
SEMIP	Secretaría de Minas e Industria Paraestatal (Ministry of Mines and Industry)
SUTERM	Sindicato Unico de los Trabajadores de Electricidad de la República Mexicana (National Electrical Workers' Union)
SUTIN	Sindicato Unico de Trabajadores de la Industria Nuclear (National Nuclear Industry Workers' Union)
UNAM	Universidad Nacional Autónoma de México (National Autonomous University of Mexico)

Mothers and the Mexican Antinuclear Power Movement

1 Introduction

On October 21, 1988, a group of approximately one hundred women and children from the village of Palma Sola, Veracruz, Mexico, gathered together to begin a march to protest the existence of the Laguna Verde nuclear power plant. On that day, the governor of Veracruz, Dante Delgado Rannauro, was due to arrive at the plant for an inaugural ceremony. The villagers, loosely affiliated with the Mexican antinuclear power movement, were trying to reach the governor to ask him to shut down the plant because they were convinced that nuclear technology was inherently dangerous and were fearful that Palma Sola would be annihilated if a nuclear accident were to occur. However, as the group reached the edge of town, members of the Mexican army brandished their weapons and ordered them to stop. The women explained that they were simply trying to march peacefully, but the soldiers told them that they had orders to prevent them from leaving the village. As the soldiers continued to threaten them with their weapons, many of the women and children became frightened and withdrew to the village. They were not able to continue with their march, so the inaugural ceremony to celebrate the construction of the Laguna Verde nuclear power plant was conducted without interruption.[1]

At that time, the village of Palma Sola was essentially in a state of siege: four warships were visible in the Gulf of Mexico, helicopters buzzed overhead, and jeeps full of soldiers constantly patrolled the area. The villagers were not allowed to have meetings; in fact, they were told that they could not congregate in large groups anywhere in the area. As news about the incident before the inaugural ceremony spread throughout Veracruz and on to Mexico City, other antinuclear groups expressed their outrage about the situation in Palma Sola. The Grupo de los Cien (Group of 100) and the Madres Veracruzanas—two elite antinuclear organizations—angrily demanded that the area be demilitarized.

Why were these groups so opposed to Laguna Verde, and why had the Mexican government threatened to use repression? This book attempts to answer these questions. It is about the Mexican antinuclear power movement—its origins after the Chernobyl accident, the relationship between the various antinuclear groups, and their interactions with the government.

Of course, any discussion of an antinuclear movement must be conducted within the context of the extensive literature on social movements. A growing literature in the last decade on what are called *New Social Movements* (NSMs) throughout the world has focused on the question of whether or not there really is anything "new" about these movements. According to some analysts, the novelty of the movements (which have included environmental, gay, and other grassroots groups, for example) has to do with their goals and practices, as well as with the transformations in the actors' identities as a result of their participation. Scholars such as Fernando Calderón, Alejandro Piscitelli, and José Luis Reyna have argued, moreover, that older paradigms, which focus on issues of class and on groups' attempts to influence or capture the state, can no longer account for these New Social Movements.[2] Nevertheless, the dominant paradigm in the study of social movements remains the *resource mobilization model;* this approach focuses its analysis on a movement's strategies and resources, both material and moral. Its strength lies in this type of analysis, whereas the New Social Movements literature is better at analyses of structural conditions that give rise to social movements, as well as of the movements' construction of meaning. A third framework, the *political process approach,* although not quite as ubiquitous, is also useful for the analysis of social movements. It examines the relationship between social movements and political structures.

This study attempts to analyze the Mexican antinuclear power movement based on aspects of the New Social Movement, resource mobilization, and political process paradigms. My argument is that the Mexican antinuclear power movement demonstrates certain "new" characteristics, such as a rejection of traditional party politics, a preference for internal group democracy, a search for autonomy, and a transformation of identity. Yet, old ways of doing politics are also present in the Mexican movement: it has sometimes succumbed to hierarchicalism, sexism, and classism. And perhaps most important, the Mexican state has demonstrated its age-old willingness to co-opt and coerce the movement's participants, as it has done with movements of the past.

In addition, this study analyzes a common phenomenon in Latin America—the mobilization of women based on a common identity as mothers. Many feminist scholars have argued that this tactic is futile because it reinforces the traditional sexual division of labor in society. Yet the case of the Madres Veracruzanas illustrates the complexities of using such a tactic. In my view, the Madres Veracruzanas have actually pushed back the traditional boundaries of the Mexican sexual division of labor.

The New Social Movements Literature

Any discussion of the New Social Movements literature must include a differentiation between empirical and theoretical studies of social movements. Empirical studies have focused on movements concerned with the interests of women, peasants, gays, and the environment. Moreover, in Latin America these studies have also dealt with movements in Christian base communities and with urban popular movements.[3]

Most of the literature on New Social Movements, however, has been devoted to theory. The literature on exactly what is "new" about the NSMs has grown significantly. Although not all scholars agree on the various details, certain trends can be identified as new. Many scholars have either explicitly or implicitly compared the new with the "old" movements, which were often based on labor or working-class issues. Alan Scott, for example, has developed a typology of four categories to contrast the workers' movements with NSMs: "location, aims, organization, and medium of action innovation."[4] He argues that workers' movements have been located within the polity, whereas NSMs are situated solidly in civil society. The goals of workers' movements often involve political integration and economic rights, but the goals of NSMs tend to involve a defense of civil society and changes in values and lifestyles. Further, NSM organizations tend to be small, grassroots groups, whereas workers' groups are more formal and hierarchical. Finally, according to Scott, NSMs usually take direct action and utilize cultural tactics, whereas the "old" movements engaged in the more traditional strategies of political mobilization.[5]

Moreover, Alberto Melucci has argued that NSMs' issue areas exist alongside those of "traditional social groupings (such as classes, interest groups, and associations)."[6] Yet grassroots groups associated with New Social Movements operate differently. They are "dispersed, fragmented and submerged in everyday life"[7] and practices (e.g., housewives who mobilize based on their own and their children's concerns). Moreover, individuals' participation in these groups fluctuates over time, with intense mobilization occurring during times of crisis; members then often withdraw into what Melucci has termed "submerged networks" often based on friendship and other everyday ties and concerns.[8] Yet, "when small groups emerge in order to visibly confront the political authorities on specific issues, they indicate to the rest of society the existence of a systemic problem and the possibility of meaningful alternatives."[9]

Melucci also argues that NSMs are "self-referential" in that the groups associated with them not only have specific, external goals, but also value their participation as an end in itself. The movements represent a "symbolic challenge to the dominant code."[10] Thus, the members of environmental groups put their beliefs into practice by attempting to live in harmony with the environment, or the members of women's groups similarly may try to challenge the subordination of women in their everyday lives.

Yet another characteristic often cited in the NSM literature is the groups' disdain for traditional politics and their attempts to keep the state at arm's length. Calderón, Piscitelli, and Reyna have argued that new movements in Latin America have looked askance at the state, political parties, and traditional leaders such as *caudillos*.[11] These movements actually question the role of the state in politics and in the economic realm. Calderón, Piscitelli, and Reyna claim that the movements are the products of crises provoked by states' projects. A few decades ago, states were deeply involved in fomenting industrialization and education. These projects—along with expanded urbanization, which brings together different sectors of the national population—have laid the groundwork for the emergence of these movements. "Essentially, there is a terrible tension between society and the state, which necessarily entails the germination of a new power structure that, for now, is more latent than manifest."[12] The NSM approach thus attempts to identify the structural factors that give rise to New Social Movements.

In addition to contrasting the new and the old movements, certain scholars have attempted to explore the inner dynamics of the new movements. Melucci, for example, has developed the concepts of *collective identity* and *submerged networks*. The actors in new movements construct a collective identity—"a movable definition of themselves and their social world, a more or less shared and dynamic understanding of the goals of their action as well as the social field of possibilities and limits within which their actions take place."[13] The idea of submerged networks is based on Melucci's observation that many environmental and women's movements, for example, are tied together not by formal hierarchical structures but rather by informal grassroots groups. He argues that these groups experience cycles of more and less intense mobilization; yet, even during times of "latency" the members of these groups are not truly beaten or inactive, given that their affective ties and everyday concerns keep them in contact with each other. For Melucci, "collective action is never based solely on cost-benefit calculations, and a collective identity is never entirely negotiable."[14]

A final distinction between so-called old and new social movements involves the actors participating in these groups. The members of the new movements tend to be "small" rather than "grand" social actors; that is, participants in old movements are peasants and workers, whereas participants in new movements frequently are housewives or students, for example. Historically, peasants and workers have been involved in social upheavals that have transformed societies throughout the world. The same cannot be said of housewives and students.

Although the New Social Movements literature makes significant contributions to the study of social movements, it also has its problems. First, it was developed in Western Europe and the United States, and although it may be extremely useful in those contexts, it is less so in the Latin American context. It does not seem relevant in this particular context to claim that issues of class—struggles between capital and labor—are no longer relevant or primary. Second, the NSM literature can be faulted for focusing its analysis on the construction of identity and meaning while excluding the development of strategy and organization. As Bert Klandermans and Sidney Tarrow have argued:

> Resource mobilization theory has been criticized for focusing too much on organization, politics, and resources while neglecting the structural preconditions of movements—that is, for focusing too much on the "how" of social movements and not enough on the "why." . . . The new social movement approach has stimulated the opposite criticisms.

Some contend that it focuses in a reductionist way on the structural origins of strain and does not pay enough attention to the "how" of mobilization.[15]

Sonia Alvarez and Arturo Escobar have argued that, given these strengths and weaknesses, it is best to combine the approaches when analyzing Latin American social movements. They argue that "a more disaggregated and nuanced view of contemporary Latin American social movements and social change depends on a careful blending of the two prevailing theoretical and methodological approaches. . . . Such a view requires us to explore the nexus of the institutional and the extrainstitutional, the 'old' and the 'new,' the cultural and the political, the local and the global, the modern, the premodern, and the postmodern."[16]

The Resource Mobilization Model

The dominant approach in the study of social movements is the resource mobilization model, which is often associated with elite theories in the study of U.S. politics.[17] This approach maintains that social movements are political entities that emerge when aggrieved actors band together to make particular political claims. It assumes that grievances are ever present but that movements emerge when participants have access to sufficient resources to maintain collective activity. Resources can be material or moral and "can include legitimacy, money, facilities, and labor."[18]

Doug McAdam has identified four strengths in this approach to the study of social movements. First, resource mobilization has served as a very useful alternative to the *classical model* (usually associated with the pluralist perspective in U.S. politics), which views social movements as expressions of psychological problems.[19] The resource mobilization approach instead treats social movements as political phenomena. The second strength of the model is that it is assumed that movement participants are making rational calculations in their choice to join a social movement rather than to pursue conventional political avenues in the system. A third strength concerns the fact that resource mobilization theorists have also looked to *external* sources of support that can aid a social movement. External factors, not just actors' internal perceptions of their grievances, can lead to the emergence of a social movement. Fourth, resource mobilization theorists, unlike classical theorists, do not see social movements as mere short-lived eruptions of protest activity; instead, they believe that the life of such movements will potentially be long if resources are available to keep the participants together and afloat.[20]

The resource mobilization approach has also been heavily criticized, however. It relies too heavily on the idea that resources are critical to the emergence and development of a social movement, yet its definition of resources is quite broad and vague. The model also lacks a theoretical connection between objective and subjective conditions for mobilization. That is, do aggrieved parties with available resources *always* form a movement? The model does not treat subjective conditions as problematic in any way. Finally, resource mobilization theorists often assume that elites play an important role in the formation of movements: they represent an important "resource." Yet, as McAdam has pointed out, this assumption is not borne out empirically; in many instances, elites have fostered the decline rather than the sustainability of social movements.

The Mexican Political System: Corporatist and Authoritarian

Mexico's current political system has its roots in the period immediately after the Mexican Revolution of 1910. During the 1920s, Plutarco Elías Calles founded the party that would come to rule Mexico for the rest of the century. In bringing together various revolutionary leaders, he institutionalized the representation of the various groups and classes that had participated in the revolution. Lázaro Cárdenas, president between 1934 and 1940, subsequently gave the Partido Nacional Revolucionario (now the Partido Revolucionario Institucional [PRI]) its structure, based on separate sectors for peasants, labor, the military, and "popular" groups.[21]

Many analysts agree that Mexico's political structures are corporatist. Ruth Berins Collier and David Collier argue that corporatism includes "(1) state structuring of groups that produces a system of officially sanctioned, non-competitive, compulsory interest association; (2) state subsidy of these groups; and (3) state imposed constraints on demand-making, leadership, and internal governance."[22] The Mexican system includes all of these characteristics: the PRI mobilizes the population based on its corporatist structure, and until the recent strengthening of the opposition parties—the Partido de la Revolución Democrática (PRD) and the Partido de Acción Nacional (PAN)—mobilization outside the boundaries of the PRI was extremely difficult.

Although a corporate political structure has been evident in several Latin American countries at different times throughout this century, Mexico's version has been the most enduring. Authors such as Alfred Stepan and John Sloan have argued that corporatist regimes were put into place in Brazil from 1937 to 1945 and in 1964, as well as in Chile after the coup of 1973, and in Peru between 1968 and 1975.[23] Yet none of these regimes was as long-lasting as the Mexican system.

Certain characteristics of the Mexican political system have contributed to its durability. Analysts argue that although Mexican political structures are authoritarian, the overall system is nonetheless inclusionary and not as repressive as many other authoritarian regimes. Political structures within the PRI and the state bureaucracy engage in "top-down" relations with civil society, but the system "encourages individuals and small groups to solicit favors (distributive benefits) from a paternalistic bureaucracy and party."[24] The system is also inclusionary in that leaders of movements are often absorbed and co-opted by the political system: these leaders and the groups they belong to are acceptable as long as "the goal is the incorporation of the disadvantaged group into the existing distribution framework rather than the destruction of the system."[25] The Mexican political system remains authoritarian, however, and will use violence and repression when necessary. The Tlatelolco massacre of 1968, in which hundreds of students were killed by police and military personnel, serves to remind dissidents that violence is always a possible response when the political structure is challenged.

Democracy and Mexican Social Movements

Though the literature on New Social Movements emerged from a postindustrial context in the United States and Western Europe, many scholars of Latin American social movements have embraced the New Social Movements approach. For these scholars, it has been refreshing to observe movements challenging authoritarianism and pushing for democracy after years of authoritarian rule. Indeed, one scholar of Latin American social movements notes that "the enthusiasm that new social movements is producing has not seen daylight since dependency theory."[26]

Of course, one of the central tenets of this paradigm is that social movements of the last two or three decades are "new" and different compared to movements of the more distant past. Regarding Mexico, some scholars argue that popular movements after the 1968 massacre at Tlatelolco have several distinctive characteristics. One difference is that in the contemporary period the sheer number of movements is a new phenomenon. These range from urban movements in poor neighborhoods to the Zapotec movement—the Coalición de Obreros, Campesinos, y Estudiantes del Istmo (the Coalition of Workers, Peasants, and Students of the Isthmus [COCEI]) in Juchitán, Oaxaca. A second difference can be found in the fact that women now play a much larger role in the movements (though not necessarily at the leadership level). Moreover, the perception of a distinction between the public and the private in Mexican society may be breaking down.[27] Finally, these movements are purportedly different because their demands include a defense of rights—whether land, labor, or human. The tone in which the demands are made is also different: the participants no longer accept the government's long-term authoritarian practices and instead insist that the Constitution be respected.[28]

What about mobilization before 1968, however? During the period immediately after the Mexican Revolution and for several decades thereafter, Mexico experienced consistent economic growth based on an *import substitution model* of economic development. Despite this fact, however, the high rate of rural-to-urban migration meant that not enough jobs were created, so the number of urban poor also grew.[29] Before 1960, the Mexican state—from the point of view of the proponents of New Social Movements—was able to manage societal pressure through its institutions. Specifically, the PRI, the state party, absorbed demands through its three sectors: the Confederación de Trabajadores Mexicanos, the Confederación Nacional Campesina, and the Confederación Nacional de Organizaciones Populares (CNOP).

This way of doing things changed, however, as the gap between rich and poor grew and as a student movement demanded greater democracy in 1968. The government responded with brutal repression at the Plaza of Three Cultures (Tlatelolco) in Mexico City on October 2, 1968. As many as three hundred students and other demonstrators were killed by the fire from army tanks, helicopters, and police sharpshooters. Despite this incident of repression, popular mobilization did not come to an end, as members demanded that the government address their grievances—mostly the need for urban services, jobs, and democracy in general. During the 1970s, urban popular movements in places such as Durango, Monterrey, and Juchitán demanded that the government respond to the specific needs of the people, which included a lack of water, land, and housing. In many of these movements, students provided "leadership, ideology, and structure."[30]

The year 1982 marked another important turning point for the urban poor and popular movements. During the 1970s, Mexico had discovered large oil reserves, which had raised high economic expectations in Mexico and abroad. Foreign banks offered Mexico generous loans, and President José López Portillo accepted them, assuming that Mexico would have little difficulty repaying them with the country's large oil reserves. No one had anticipated that the world price of oil would collapse, however. After the collapse, Mexico found itself unable to make the payments on the loans: it was now one of the world's biggest debtor nations.

The new president, Miguel de la Madrid, took office in 1982 and began to institute neoliberal economic policies in order to bring order to the economy. In addition, the International Monetary Fund advised that social spending be curbed. As a result, the government now had fewer resources to concede to popular movements.

Nevertheless, the mobilization among popular organizations continued. After the 1985 earthquake in Mexico City, for example, the government was noticeably ineffective in responding to the crisis. Popular organizations responded quite well, however. The First General Assembly of Neighborhoods responded to housing problems by generating a list of people who still needed housing two years after the earthquake.[31] And a group of garment workers—who lost their jobs after the earthquake destroyed four hundred centers of production—chose to form a union in order to defend their interests.

But have these movements been successful in pushing for democracy, as the proponents of the New Social Movements approach claim? Certainly many scholars of recent social movements in Southern Cone countries have argued that these organizations are at least partially responsible for the turn to democracy because they not only push the state toward democratic decision making but also engage in democratic practices within themselves.[32]

Although most scholars engaged in the study of Mexican social movements have been more cautious in claiming that popular movements in Mexico have been successful in pushing for democracy, Joe Foweraker, in his study of the teachers' movement, has argued that popular movements of the last two decades "challenge the exclusionary policies of the PRI government in particular, and . . . the clientelistic patterns of political control in general. They are both 'institutionalist' *and* nonconformist."[33] On the other hand, Judith Hellman and Paul Haber take exception to this argument. Hellman believes that "the development of popular movements represents a less dramatic and significant departure from clientelism than Foweraker and other analysts have appreciated. I see these movements as deeply enmeshed in clientelistic patterns from which they escape only very rarely."[34] Nor is Haber convinced that popular movements have pushed the process of democratization further in Mexico: "It is far from clear to what extent popular movements have actually succeeded in forcing a wide political opening and whether events that have transpired since 1989 have succeeded in pulling Mexico back from the brink of becoming more democratic. The character and results of the August 1994 elections in Mexico have not been widely perceived by popular movements as major steps in a more democratic direction."[35]

This study, then, seeks to analyze the Mexican antinuclear power movement within the context of the literatures on the New Social Movements and the process of "redemocratization" in Latin America. Specifically, I seek to answer two questions. First, can the Mexican antinuclear movement be considered a New Social Movement? Certainly the antinuclear movement at first glance appears to qualify as new when compared to other contemporary popular movements in Mexico. The movement against the Laguna Verde nuclear power plant was (is) made up of organizations representing Mexicans from diverse class backgrounds; moreover, the movement's goal—to rid Mexico of nuclear energy—appears to match the NSM definition, which includes the pursuit of postmaterial goals.[36]

Second, has the antinuclear movement pushed the Mexican government to be more democratic? Here I am using Jonathan Fox's definition of democracy: "free and fair electoral contestation for governing offices based on universal suffrage, guaranteed freedom of association and expression, accountability through the rule of law, and civilian control of the military."[37] In addition, a democracy demonstrates "respect for associational autonomy, which allows citizens to organize in defense of their own interests and identities without fear of external intervention or punishment."[38]

This book is divided into ten chapters. Chapters 2 through 7 recount the development of the movement as the tactics of the antinuclear groups evolved to meet changing circumstances. In the early years of the movement, 1986 to 1988, the groups mobilized thousands of people, and protests occurred on a regular basis. However, as it became clear that the government intended to operate the plant despite the protests and often used repression against the movement activists, fewer people were willing to take to the streets. Thus, beginning in October 1988, when the Laguna Verde plant went on-line, the antinuclear groups changed their tactics and started to engage in direct lobbying of the government. They also began to focus their opposition on particular aspects of the Laguna Verde problem, such as the flawed emergency evacuation procedure. Chapter 8 describes the decline of the movement. As it became clear that the government was not going to give in easily to movement demands, some participants dropped out, others were co-opted, and another group ceased to participate because of fear of government repression. Chapter 9 analyzes the role of class and gender in the development of the movement. The role of gender—specifically, mobilization based on an identity of motherhood—has been especially important in the participation of the Madres Veracruzanas in the antinuclear movement. Finally, chapter 10 is devoted to an examination of the Mexican antinuclear movement in light of various theoretical approaches, including the New Social Movements and the political process approaches.

2 The Origins of the Mexican Antinuclear Movement

This chapter traces the origins of the Mexican antinuclear power movement. In the early 1970s construction began on a nuclear power plant at Laguna Verde in the state of Veracruz. At first, few Mexican citizens were even aware of its existence. Residents of the state of Veracruz—if they had an opinion at all—generally were pleased that such a high-tech plant was being built in their state. These views changed with the occurrence of the Three Mile Island and Chernobyl accidents. When the Mexican media gave extensive coverage to the Chernobyl accident of 1986, the people of Veracruz became extremely concerned about the Laguna Verde plant, which was in its final stage of construction. Groups such as the Madres Veracruzanas emerged specifically to address the nuclear issue. The Madres were joined by already-existing environmental groups, such as the Grupo de los Cien and the Pacto de Grupos Ecologistas (Pact of Environmental Groups), and these groups now focused their attention on the nuclear power plant at Laguna Verde.

Mexico's Nuclear Energy Program

Mexico's nuclear energy program began in 1965, during the administration of President Gustavo Díaz Ordaz. During the late 1960s and early 1970s, the Mexican government was enthusiastic about nuclear power for several reasons. Mexico had not yet discovered its largest oil fields, and the government was preoccupied with finding alternative sources of energy. In addition, nuclear power experts predicted that nuclear energy would be cheaper than energy produced from other sources. Finally, Mexican government officials were anxious to keep up with advancing technology.[1]

The government's first attempt to embrace nuclear technology occurred in 1979 in the state of Michoacán on Lake Pátzcuaro. Government officials planned to sponsor a nuclear reactor—purely for research purposes—in the small Indian town of Santa Fé, on the north shore of Lake Pátzcuaro.[2] A nuclear workers' labor union, Sindicato Unico de Trabajadores de la Industria Nuclear (SUTIN), was at the forefront of the proposed project. One of the founders of the union had purchased a vacation home in the area and thought that Lake Pátzcuaro would be an ideal site for a nuclear laboratory. To that end, members of the union began lobbying the local population—the Purépecha people—to support the reactor. The Purépecha were told that the laboratory would be a source of new jobs for the local population, so they were at first enthusiastic about the project but later changed their minds as a result of the antinuclear arguments presented by environmentalists such as José Arias Chávez and Carmen Buerba.

By 1980 the government had not officially announced the nuclear project, but the union continued to give indications that the laboratory would indeed be built. At that point, however, Purépecha leaders denounced the project at a worldwide conference of indigenous peoples held in Rotterdam. Throughout 1980 and 1981 more Purépecha began to mobilize against the nuclear laboratory and were supported by the Asociación de Tecnología Apropiada (Association for Appropriate Technology [ATA]), which included intellectuals from Mexico City as well as from the Universidad de Michoacán. After a series of protests and meetings with government officials, the government canceled the proposed Centro de Ingeniería de Reactores (Center for Reactor Engineering [CIR]) on May 26, 1981.[3] It did not explain the cancellation, but apparently the depth of the local population's hostility contributed to the decision.

Originally, the Mexican government had planned to construct a nuclear plant on the U.S.-Mexican border—as a joint venture—in order to provide electricity to Arizona, California, Sonora, and Baja California. The site of the proposed plant was later changed to Laguna Verde, Veracruz, after the Mexican government consulted with the Instituto Nacional de Energía Nuclear, the Banco de México, and the Stanford Research Institute, a North American consulting firm. After a bidding process in 1972, the Mexican government awarded General Electric a contract to build two nuclear reactors (Mark II BWR5) to be contained in a plant at Laguna Verde. Throughout the plant's history, other companies from various countries have also been involved in aspects of its construction: Bufete Industrial, Burns and Roe, and Bechtel.[4] Overall, the Laguna Verde plant took twenty-three years to build: four presidential administrations as well as some forty companies and groups were involved.

The Environmental Movement in Mexico

Mexico had had environmental problems for decades, but the emergence of an environmental movement was a relatively recent phenomenon in the early 1970s.[5] Mexico's rapid urban growth since the 1940s had led to inadequate waste disposal, water scarcity, and pollution of air, soil, and groundwater.[6] In addition, the petroleum industry had destroyed entire ecosystems along the Gulf Coast, thus damaging Mexico's fishing industry. In the country's northern region, the heavy use of pesticides and irrigation with salinated water had reduced the land's productivity.[7]

Despite these problems, no independent environmental movement emerged until the 1980s. Environmental mobilization in Mexico can be traced to the government's Federal Law to Prevent and Control Environmental Pollution, which was passed in 1971 and called for the protection of Mexico's air, water, and soil. Unfortunately, the law was vague and did not include precise standards for environmental regulation and enforcement.

During the early 1970s, very few government officials were concerned about environmental issues. Their attitude began to change, however, as citizens became more affected by environmental degradation. The growing environmental problems led academics, intellectuals, and a few professional groups to raise questions about how best to protect the environment.

Environmental issues first emerged in the political arena in 1982 during Miguel de la Madrid's presidential campaign. Shortly after assuming the presidency, de la Madrid created a new cabinet-level ministry—the Secretaría de Desarrollo Urbano y Ecología (Ministry for Urban Development and the Environment [SEDUE]). He also embarked on what Stephen P. Mumme has described as "preemptive reform"[8] when he and members of his administration recognized that the environment was a growing political issue and that their goal would be to channel that mobilization in a particular direction. In 1983–84 the government sponsored a series of public meetings at the local, state, and national level. These meetings served to heighten citizens' awareness of environmental problems and to familiarize them with new environmental legislation. From June 4 to June 8, 1984, the government sponsored a national environmental congress, which President de la Madrid attended and which "had the intended effect of promoting the organization and development of existing and new environmental interest groups and legitimizing their participation in Mexican politics."[9] Indeed, during this period organizations such as the Movimiento Ecologista Mexicano (MEM), the Alianza Ecologista (AE), and the Pacto de Ecologistas were formed. Although these organizations were ostensibly national and multiclass, their members were mostly from Mexico City and tended to come from middle-class backgrounds. These three groups would play important roles when the mobilization against Laguna Verde began a few years later.

The Antinuclear Movement in Mexico

The antinuclear movement began in 1986 in response to two events: the accident at Chernobyl in April 1986 and President Miguel de la Madrid's announcement on September 1, 1986, that the Laguna Verde project would be completed and the plant would be put into operation. In 1986 and 1987 numerous groups (the Madres Veracruzanas, for example) emerged in response to these events. Other, already-existing organizations (Lions Clubs, labor unions, and the Catholic Church) now included opposition to Laguna Verde in their activities. The members of the groups actively working against nuclear energy in Mexico ranged in socioeconomic class from Mexico's most prominent artists and intellectuals in the Grupo de los Cien to peasants in local organizations.

The most intense antinuclear opposition emerged in the state of Veracruz, the area that would be most affected by a nuclear accident. People from every social class spoke out against the plant. Hotel owners, for example, were worried that the plant would hurt tourism. Catholic bishops in Veracruz used the pulpit to express their opposition to nuclear power, arguing that the Laguna Verde plant would intensify, rather than solve, Mexico's economic problems. Poor peasants and wealthy cattlemen joined the movement because they feared that their crops and animals would be contaminated by radiation.[10]

The antinuclear groups expressed their opposition to Laguna Verde for numerous reasons—many of the same arguments used by activists in the United States and Europe.[11] First, the plant's technology was already obsolete and poorly designed. Moreover, one of the two vessels sent to the plant had been damaged shortly after its arrival, making Laguna Verde especially susceptible to an accident. Second, the plant was built near a fault line and a volcano, which increased the chance of an accident. Third, even if the nuclear power plant were to operate "normally," it would emit low levels of radiation that could harm living beings and the environment. Fourth, there was no safe way to dispose of nuclear wastes. In addition, the villagers of Palma Sola discovered that the emergency evacuation plan for the Laguna Verde region was deeply flawed.[12] Finally, the plant's opponents claimed, the electricity produced by Laguna Verde would be twice as expensive as electricity produced by conventional plants.

The antinuclear activists used a variety of tactics to express their opposition to the plant. In Veracruz, they staged numerous marches, many of which ended in Xalapa, the state capital, where opponents stood before the Governor's Palace demanding that the plant be closed. The cattlemen of the village of Palma Sola organized three traffic blockades on Mexico's main Gulf Coast highway, which cuts through the town. One blockade—set up in June 1988—lasted three days and was finally broken up by state and federal military forces. In the cities of Córdoba and Xalapa, residents periodically organized voluntary blackouts in the evenings as dramatic gestures to express their opposition. Many residents of Veracruz scribbled notes of protest on their electric bills, and some threatened to withhold payments until Mexico abandoned its nuclear energy program.

Since 1990, however, the movement has been unable to attract large numbers of people. Instead, the various antinuclear groups have adopted new strategies. The Madres Veracruzanas have met with top government officials, and the Coordinadora Nacional Contra Laguna Verde (CONCLAVE) has attempted a large legal campaign to prevent the plant from operating.

However, certain sectors of Mexican society support Mexico's nuclear power program. The most vocal of these groups has been the Sindicato Unico de los Trabajadores de Electricidad de la República Mexicana (SUTERM), Mexico's electrical workers' union, which has threatened a national blackout if the Mexican government suspends the nuclear program. In addition, the Comisión Federal de Electricidad (CFE) supports nuclear energy. The most important support of all comes from Mexico's dominant political party, the PRI, whose officials have portrayed the opponents of nuclear energy as backward enemies of technological progress—a portrayal that has been a detriment to the antinuclear movement.

The Organizations

The following are the most active groups in the antinuclear movement.

The Grupo de los Cien

Made up of intellectuals and artists who decided to join together in order to voice their opinions on Mexico's environmental problems, the Grupo de los Cien has been a key organization in the antinuclear coalition.[13] Its main focus is environmentalism in general: the nuclear issue is only one of its many concerns. Within the Grupo, three people are especially active—actress Ofelia Medina, writer Homero Aridjis, and artist Feliciano Béjar. Not all intellectuals in Mexico have supported the Grupo, however. For example, Octavio Paz—a Nobel Prize winner in literature—was in favor of nuclear energy.

The Grupo's strategy has been similar to that of the other antinuclear groups. Whenever a particular crisis erupts at the Laguna Verde plant, the group uses the media to publicize the problem in order to mobilize the population at large. Often, this strategy means buying advertising space in Mexico City newspapers. This strategy has been limited to the print media because a heavy-handed censorship has been imposed on Mexico's television stations regarding the nuclear issue. (Most of the television stations are actually owned by the government.) Feliciano Béjar argues that their ability to publish in the newspapers is also limited because the group does not have enough money to buy much advertising.[14]

CONCLAVE

The Coordinadora Nacional Contra Laguna Verde (National Coordinating Committee against Laguna Verde [CONCLAVE]) is an umbrella group linking various small antinuclear organizations. This group, like the Madres

Veracruzanas, is composed of various chapters: the most active chapters are in Xalapa and Mexico City. CONCLAVE's members are, for the most part, highly educated professionals, scientists, and academics. Their tactics have ranged from participating in demonstrations to mounting legal challenges to the presence of the Laguna Verde plant. During the 1990s these activists mounted a legal campaign, filing injunctions in various courts in an attempt to have the plant closed down. In addition, they have attended environmental conferences, including the environmental summit at Rio de Janeiro.

The Villagers of Palma Sola, Veracruz

The people of Palma Sola can be considered one entity, although officially their participation was led and mediated by the local Cattlemen's Association, the Asociación de Ganaderos, which is affiliated with the state-wide cattlemen's organization. Palma Sola is an agricultural village located a mere two kilometers from the Laguna Verde plant; the proximity of the residents to the plant led many of them to believe they would be in grave danger in the event of a nuclear accident. The cattlemen also worried that their beef and milk products would be rejected by consumers concerned about radioactivity. During the early years of the movement, the people of Palma Sola worked tirelessly against the plant, helping to organize three blockades of traffic and traveling to Mexico City to meet with politicians. They worked in conjunction with the Madres Veracruzanas from 1986 through early 1989.

The Comité Antinuclear de Madres Veracruzanas

The Comité Antinuclear de Madres Veracruzanas was founded in early 1986 to educate mothers throughout the state about the danger that nuclear energy poses for children and for the population at large. The mothers' organization has chapters in the port of Veracruz, in Xalapa, and in Córdoba, Veracruz. Approximately two hundred women belong to the organization, with a core group of twenty-five participants. The Xalapa chapter, the largest and most active of the three, was founded in February 1986 by several concerned women. In the beginning these women, the majority of them from upper-middle-class backgrounds, educated themselves about nuclear technology. They persuaded a local environmental-ist, Thomas Berlin, to give them private classes on nuclear technology. Then, armed with this knowledge, they proceeded to organize protest activities against Laguna Verde.

The members of the Madres Veracruzanas define themselves only as mothers defending their children from imminent harm. They argue that their group is *not* political. When asked how their group can be apolitical, given that they have to use the political process in order to achieve their goal, they respond that the political sphere is dirty, corrupt, and male, so they want no part of it but have been forced to participate in it to protect their children.

The mothers' protest activities reflect their status as upper-middle-class women and mothers. Consciously copying the tactics of the Madres de la Plaza de Mayo of Argentina, the Madres Veracruzanas have staged protests every Saturday morning in Xalapa's main plaza since 1986. The members maintain a dignified profile. They wear white, feminine cloth-ing. While holding their protest signs, they explain their opposition to Laguna Verde to interested passersby. The women have adopted red rib-bons as their symbol and have persuaded residents and merchants through-out Xalapa to display red ribbons in their homes and businesses.

Although only one of the members had any previous political experi-ence, the women have been quite adept in the political arena. Representa-tives of the organization have managed to arrange meetings with such high-ranking politicians as Manuel Bartlett, Manuel Camacho Solís, and even former president Carlos Salinas de Gortari, who refused to meet with any other antinuclear group. The Madres have not met with Presi-dent Ernesto Zedillo, however, despite numerous requests for meetings.

The Antinuclear Groups' Philosophies

Though the many antinuclear groups agreed on one basic point—that Laguna Verde should not operate—they did not all share the same phi-losophies. Members of some of the groups—such as the Pacto de Grupos Ecologistas (many of whom later joined CONCLAVE) and the Grupo de los Cien—considered themselves to be environmentalists and had a very so-phisticated analysis of the role of high technology, the economy, and the state in the development of Mexico's nuclear program. They worried about the relationship between high technology and authoritarianism in the case of Laguna Verde. However, they were also concerned with all environmental issues in Mexico.

The Madres Veracruzanas, in contrast, were more focused on closing down Laguna Verde, and not all of them considered themselves to be environmentalists. On at least one occasion, one of the members asked the other Madres if they were environmentalists *(ecologistas)* because of their views on Laguna Verde. The member asking the question did not consider herself to be an environmentalist, and at least some of the other Madres shared her position. Thus, unlike the members of the Grupo de los Cien, most of the Madres did not hold environmental beliefs beyond their dislike of nuclear technology.

Other participants in the movement, such as the cattlemen of the village of Palma Sola, also did not consider themselves to be environmentalists. They joined the movement for economic reasons—because they feared their agricultural products would be rejected by Mexican consumers. Like the Madres, they were not particularly concerned with larger environmental issues. Instead, they tried to find concrete solutions to the Laguna Verde problem; at one point during their participation in the movement they even offered the government financial help in converting Laguna Verde into a conventional gas-fired plant.

Some antinuclear groups represented individual towns and villages in the state of Veracruz, such as Coatepec, Misantla, Banderilla, Martínez de la Torre, and Emilio Carranza. Most of the members of these groups were not highly educated and did not have access to important government officials (as did the Madres Veracruzanas). Deeply moved by the extensive coverage in Mexico of the Chernobyl accident, the members of these groups were primarily motivated by fear: they were convinced that their homes and towns would be destroyed if Laguna Verde were to operate. They did not, however, have a sophisticated understanding of the relationship between economic, political, and technological factors at play in the Laguna Verde project; rather, they were simply trying to protect their homes and towns.

Individual Activists

Certain individuals came to play a major role in the antinuclear movement. These activists include Pedro Lizárraga, José Arias Chávez, Marco Antonio Martínez Negrete, and Thomas Berlin. All of these leaders were affiliated with Mexican universities. Lizárraga is a psychology professor at the Universidad Veracruzana. An engineering professor at the Universidad Nacional Autónoma de Mexico (UNAM) in Mexico City, Arias Chávez began his participation in environmental issues in 1967 when he helped to found the group that is now called the Fundación de Desarrollo (Foundation for Development). He was also a member of the Association for Appropriate Technology, which helped to block the construction of a nuclear laboratory in Michoacán. During the 1980s, Arias Chávez was a leader of both the ATA and the Pacto de Grupos Ecologistas, which later joined CONCLAVE.

Marco Antonio Martínez Negrete holds a doctorate in physics and, like Arias Chávez, teaches at UNAM. He too participated in the antinuclear struggle in Michoacán and was a member of the Pacto de Grupos Ecologistas and later of CONCLAVE.

Thomas Berlin, by contrast, teaches architecture at the Universidad Iberoamericana in Mexico City. He holds a doctorate in regional planning at the University of Stuttgart and has worked as an energy consultant for the Veracruz state legislature.

The Madres Veracruzanas in Their Own Words

This study focuses primarily on the antinuclear activity of the Madres Veracruzanas. Though the Madres were not the largest or most militant group in the antinuclear movement, they became the most important. Why? First, they never wavered in their antinuclear stance. Although other groups had folded by the early 1990s, the Madres Veracruzanas were still active in 1998. Second, they were always at the center of antinuclear activity and, moreover, were the only antinuclear activists who had access to the president of Mexico (during the Salinas administration). No other groups had such easy access to the highest levels of government. Finally, a special study of this group is important because it illustrates the gendered dimension of the antinuclear movement's activity and the state's response.

The women who formed the Madres Veracruzanas in 1986 decided they needed a group of their own even though other environmental groups (such as the Pacto de Grupos Ecologistas and the Movimiento Ecologista Mexicano) were already engaged in antinuclear activity. In a group interview conducted on July 22, 1988, Claudia Gutiérrez (psychologist), Adela Chacón (schoolteacher), Rebeka Dyer (school administrator), Mercedes Solé (linguist), and Irma Landa (physician) recounted why and how they decided to form the Madres Veracruzanas. Like all of the interviews I conducted in Mexico, this interview was in Spanish, which I have translated into English.

VGG: How did your group get started?

REBEKA: In September 1986 I published a letter against nuclear power, saying that Mexico should think about how nuclear accidents harm children. This was after accumulating some knowledge about nuclear energy.

ADELA: The news in the paper worried me, especially the accidents at Three Mile Island and Chernobyl. My knowledge of Laguna Verde was vague. No one gave the plant much thought. In fact, earlier, if anyone thought about Laguna Verde at all, they thought it was a great idea. [After reading Rebeka Dyer's letter in the newspaper, Adela called her.]

REBEKA: Fifteen years ago there were some criticisms of the plant, but these mostly came from university professors and students only. Eight years ago a Chilean woman also raised criticisms, but no one paid any attention.

ADELA: I was against nuclear energy, but I knew that I couldn't do anything by myself, so I called Rebeka. At first, our only goal was to inform the public about the dangers of nuclear energy. So we called a meeting for mothers concerned about the Laguna Verde plant.

VGG: Why only women?

REBEKA: We didn't know what to do at first. We saw that other groups already existed; for example, there were demonstrations in the Parque Juárez [in Xalapa] in 1986. We were aware of the Niño Ecologista—a ten-year-old who, along with his father, walked all the way from Oaxaca to Veracruz to protest the plant. Yet I didn't feel comfortable about

joining the already-existing groups because of my work and family schedule.

CLAUDIA: In February 1986 we decided to call a meeting for mothers; we brought our friends. I invited Antonio Bretón [one of the cattlemen from Palma Sola]; I knew him through our children at school. But
no club in town would lend us their hall. So finally we managed to land
the Casino Español. There I met Consuelo Landa and also Letti Tarrago
[a well-known artist], who volunteered to create posters for the group.

REBEKA: The problem was that we didn't know *whom* we were
fighting against. The government has different faces. My husband warned
me to be careful; he was a student in Mexico City during the 1968
Tlatelolco massacre. We're not in this because we want political positions or jobs.

MARGARITA: Men don't participate until they know they're going to
win. Women are different: they're participating because they want to
protect life above all, their children.

REBEKA: As for the movement in general, we have to figure out what
the government thinks. Many of the members of the movement are
scared of the government; they still remember 1968. But we see nuclear
power as even more frightening than any possible repression on the
part of the government.

MARGARITA: Men don't have the experience of giving birth. I was
antinuclear, but only at home, so I finally decided to join the movement.

REBEKA: My family became jealous because I spent so much time on
the movement. What's interesting is that in Mexico normally it's the
husbands who speak with important people. But now the Madres
Veracruzanas are the ones meeting with the politicians.

VGG: What about sex and class as factors? [The women all agreed
that their husband's connections and being members of the upper middle
class have helped their cause as Madres Veracruzanas. The newspapers
are also more likely to cover their activities because of their privileged
backgrounds.]

LETICIA: Though we're of middle-class background, we've decided
to join a movement instead of playing canasta, as many upper-middle-

class women do in Mexico. We spent three days learning about nuclear energy from Thomas Berlin.

MERCEDES: I remember seeing pictures of antinuclear women in England chained to fences. Mexican women are not like the Europeans. The Madres would never chain themselves to anything; we would lose our dignity and the respect of Mexican society, and our group would fail in its mission.

VGG: How do the Madres get along with the other groups?

MARGARITA: All of the groups feel united on the antinuclear issue. But the mothers are also fearful of other groups' other goals. The mothers are also afraid *for them.* There is so-called free expression in Mexico, but there is always the possibility of repression. And often these other groups abuse free expression. We're united on only *one point.*

MERCEDES: Yet this has been a good opportunity to meet people of different backgrounds.

REBEKA: We have begun to see more differences within the coalition—for example, voting, claims for justice, and housing. Some of their goals are too radical. We also don't identify with some of their tactics; for example, some of the other groups use obscene language in the streets during demonstrations. The mothers try to maintain a dignified posture. Other groups also tend to be less organized. We've incorporated a division of labor.

In the beginning we were also naive. For example, I have a *comadre*[15] who is a secretary to a high-level politician in the state government of Veracruz. I used to tell my comadre everything; now we're more careful with our information. We know that they have files on us at Seguridad Pública [Public Security]. Also, all of our phones are tapped. We were scared during the period when we were filing injunctions to try to prevent the plant from operating. The Madres in the port of Veracruz have had their problems. They received anonymous threats. They were afraid that their children could be kidnapped. They were afraid to wear their red ribbons for a time.

MARGARITA: Yet Xalapa is different. Women are freer in Xalapa.

The Madres began their activities simply by having meetings and inviting interested mothers in Xalapa to attend. They then encouraged businesses to display antinuclear posters and participated in debates with pro-nuclear proponents. Later, they initiated their red ribbon campaign: red ribbons became the symbol for resistance to the nuclear energy program. After the mothers began wearing red ribbons, private homes and businesses began displaying them also. Eventually, the mothers decided that they had to become more involved, so they began a letter-writing campaign. They wrote to all of the bishops of Mexico as well as to prominent artists and journalists. The vast majority of the bishops wrote back, replying that they supported the mothers in their struggle.

• • •

With the nuclear accidents at Chernobyl and Three Mile Island, Mexican environmental groups such as the Grupo de los Cien turned their attention to the nuclear question. Other groups came into existence specifically to oppose the Laguna Verde nuclear power plant. The cattlemen, for example, were worried that their livelihood would be affected by the plant's radioactivity, and the Madres Veracruzanas argued that nuclear technology would harm their children.

The Madres formed their own group for two reasons: they believed their view of the world was different from the other groups because they were motivated only by their desire to protect their children and because they felt that other groups would not understand that their family life sometimes interfered with participation. The Madres acknowledged that their families worried that the government might take repressive measures against them. Yet they believed that a combination of gender and class factors would protect them from the government.

Clearly, none of the groups emerged specifically to push for democracy in Mexico. They realized that the government had a history of using repression against social movements: the 1968 Tlatelolco massacre weighed heavily on their minds. Nevertheless, in the words of one of the Madres, "We see nuclear power as even more frightening than any possible repression on the part of the government."

3 The Mexican Antinuclear Power Movement, 1987–1988

The Government Operates the Plant Despite Opposition

During the years 1987 and 1988 Laguna Verde was in its last stage of construction; it was now up to the government to decide when and if the plant would go on-line. Thus, the immediate goal of the antinuclear movement was to persuade the government to dismantle the plant after more than two decades of construction and an investment of more than three billion dollars. Specifically, because the antinuclear organizations believed that Laguna Verde's fate would be decided by the president, they demanded meetings with the president and with the governor of Veracruz, hoping that the governor could persuade the federal government to close the plant. During these two years, they also engaged in various tactics—sometimes simultaneously—to mobilize public sentiment against nuclear technology: they blockaded major highways, participated in a statewide referendum on Laguna Verde, and tried to reveal the plant's flaws to the public.

The Blockades of Highways

By early 1988 the activists had become frustrated with the federal government's unresponsiveness to their message, so they began to consider increasing their use of civil disobedience tactics. Movement sympathizers had already participated in voluntary blackouts in Xalapa and Córdoba, and many were also marking peso bills of all denominations with antinuclear slogans to protest the plant's existence. Governor Fernando Gutiérrez Barrios, caught in the middle, exhorted the population to keep its protest activities within the confines of the law.[1] He warned that the Bank of Mexico would nullify the marked bills and that the emergence of more marked bills could hurt the economy. He also asked the activists not to blockade highways because blockades are illegal in Mexico and could cause harm to the local population.[2]

The movement activists ignored the warnings and proceeded to blockade the coastal highway at Palma Sola, Veracruz, on February 24, 1988, at 11 A.M.[3] They declared that their primary demand was to secure a meeting with the governor and subsequently with the president. Thousands of supporters descended on the village of Palma Sola to participate in the blockade. In the meantime, the Mexican government mobilized the army and navy in the area.[4] Military personnel set up inspection posts and prevented sympathizers from joining the blockade. A few hours later, a helicopter carrying representatives from the governor's office arrived to meet with movement leaders. The protesters explained that they wanted a meeting with the governor and were prepared to continue the blockade indefinitely.[5]

The blockade was lifted after thirty-two hours on February 25, 1988, at 5:30 P.M. when Governor Gutiérrez Barrios agreed to meet with the groups at Palma Sola.[6] The governor, in an apparent move to calm the populace, declared that "the government of the republic shall continue undertaking profound scientific and technical studies in order to assure the tranquility of the people of Veracruz."[7] He promised an open meeting with antinuclear groups at Palma Sola on March 2, 1988, as well as a meeting with the PRI candidate for the presidency, Carlos Salinas de Gortari.

Almost immediately, however, on February 29, another blockade went up on the road between Minatitlán and Acayucan, Veracruz. According to Juan Marín, a leader in the antinuclear movement, this blockade was erected to show support for the other antinuclear groups involved in the earlier blockade and to express people's fear of the Laguna Verde plant. Three thousand people from fourteen villages and activists from two groups—the Frente Regional Antinuclear y Ecológico and the Asociación Ecológica del Istmo—prevented approximately twenty thousand vehicles from passing along the road.[8] The local villagers decided on this method of protest even though they knew that their actions were against the law and that they could be punished.[9]

The Madres Veracruzanas explained why blockading traffic was becoming a favorite tactic for the antinuclear groups: their numerous requests for a meeting with the president had failed.[10] Meeting with the president is a right guaranteed by the Constitution, yet President Miguel de la Madrid refused to meet with any of the groups. The Madres argued that the population was anxious because the evacuation procedure, the Plan de Emergencia Radiológico Externo (PERE), was designed to aid only people who lived within a sixteen-kilometer radius of the plant.[11] The residents realized, however, based on the Chernobyl incident, that the effects of a nuclear accident can be felt for thousands of kilometers.[12] Thus, when President de la Madrid refused to listen to the people's concerns, the antinuclear groups had to resort to civil disobedience tactics such as the blockade of highways in order to get his attention.

As planned, Governor Gutiérrez Barrios met with movement representatives and townspeople after the blockade at Palma Sola. During the almost five-hour meeting the governor tried to reassure his audience that they should not become alarmed about the plant or about the democratic process.

The antinuclear activists' reaction to the meeting was generally favorable. Miriam Zavala de Ruíz of the Madres Veracruzanas of the port of Veracruz said, "Our impression was very positive; the governor, Fernando Gutiérrez Barrios, appears to agree with us that Laguna Verde represents a danger for all of the population."[13] She also announced that "Don Fernando agreed that the problem should be studied thoroughly."[14] The antinuclear groups were also buoyed by the governor's declaration that even if the plant went on-line, it could be closed later in case it became necessary. "Only death is irreversible," he said.[15] The activists were pleased with this declaration because officials from the Comisión Federal de Electricidad, which was responsible for operating the plant, had said earlier that the plant could never be shut down once it went on-line.

During the meeting with the governor, the cattlemen of central Veracruz made a proposal that showed the depth of their dislike for the Laguna Verde plant: they would be willing to contribute financially to a conversion of the nuclear plant to a thermoelectric facility.[16] They explained that they feared their beef products might become contaminated by radioactivity and were thus anxious to prevent Laguna Verde from going on-line.[17]

Despite the meeting with the governor, nothing was resolved. Several months later, the antinuclear activists held a national meeting in Mexico City, where representatives announced that they would continue their highway blockades.[18] According to Mariano López of the Grupo Arcoíris

(Rainbow Group) of Xalapa, the activists planned several marches and blockades of roads near the plant.

On June 17, 1988, the various antinuclear groups announced plans for another round of protests in opposition to the plant. The mobilizations would take place on June 18 in Palma Sola, where a rally would be held in the morning. Subsequently, certain antinuclear activists and residents were to march across the country to the Zócalo, Mexico City's main plaza.[19] On Sunday, June 19, blockades would be erected on the Veracruz–Tampico coastal highway as well as on highways near Xalapa, Coatzacoalcos, Plan Cedeño, Misantla, and Cosoleacaque.[20] Marches would also be held simultaneously in the states of San Luis Potosí and Michoacán.

On June 19, as planned, approximately two thousand people gathered to erect blockades. They represented the various antinuclear organizations and all of the major political parties except for the PRI.[21] The activists symbolically closed the plant and put antinuclear slogans around the perimeter of the facility. Simultaneously, blockades were established in Mexico City and on the Orizaba–Córdoba (Veracruz) highway. In Palma Sola the blockade began first with a mass at the local Catholic church; then cars and trucks began to park on the pavement to block traffic on the coastal highway. Cattlemen arrived with trucks carrying water, juice, and food; the activists were prepared to blockade the road indefinitely.[22] Marta Lilia Aguilar, one of the organizers, said that two thousand people had arrived in Palma Sola—fewer than expected, but "all [were] very enthusiastic."[23]

Forty-six hours later, six hundred people remained at the blockade, and there seemed to be no end to the protest in sight.[24] The protesters hailed from several states and agreed that they wanted to force a meeting with President Miguel de la Madrid. In the meantime, the blockade was causing hardships for many motorists and truckers. Raúl Rodríguez Méndez, director of the Servicio de Carga (Loading Services) at the port of Veracruz, announced that if the blockade were to last much longer, he would have to approach the antinuclear activists to allow trucks carrying perishables to pass.[25]

Governor Fernando Gutiérrez Barrios denounced the "completely illegal" actions of the protestors.[26] He asked the activists to reconsider their strategy and said that he did not want to have to intervene directly. Complaints were increasing, he said and warned that blockades are illegal and that "many of the aspects of daily life are being affected, and this is not the way to protest."[27]

Pedro Lizárraga, an antinuclear activist from Xalapa, responded to the governor through the press. He argued that "90 percent of the people of Veracruz do not want the plant. The governor has to pay attention to his people."[28] The Madres Veracruzanas also responded to the governor's criticisms, saying that they supported the blockade (many of the women were participants).[29] They also published an announcement in newspapers to clarify their position on the blockade:

> The blockade of highways has been criticized and categorized as unconstitutional; nevertheless, those who disapprove of this measure, in this case, have not thought about, or are not familiar with, the motives that led to this act. The antinuclear movement has struggled for more than two years so that the president of the republic would listen to the reasons why Veracruz does not accept the Laguna Verde nuclear plant. We have been peaceful, respectful, and patient, acting always within the framework of the law, and not even this has allowed us a direct conversation with the federal executive, Miguel de la Madrid Hurtado. How much more time within the framework of the law should we wait before the reactor is loaded . . . ? Could the blockade have been avoided? Sincerely, Comité Antinuclear de Madres Veracruzanas.[30]

Those Madres not at the blockade gathered in front of the governor's offices wearing black—as if in mourning—while the blockade continued.[31] The governor announced that he had arranged for another meeting between blockading antinuclear representatives and officials of the federal government.[32] At the meeting, he asked the activists to stop protest activities because they were affecting the peace and the economic life of the region, and reminded them that the Governor's Palace was always ready to hear their opinions and requests.[33] Blockade participants Antonio Bretón, Roberto Helier, Yolanda Rivera, Antonio Sosa, and Marta Lilia Aguilar told him that they would pass this message on to their fellow activists.[34]

Forty-eight hours after the blockade began, a new problem emerged. Truck drivers, frustrated with their inability to cross the blockade, erected their own blockade at another part of the coastal highway, near the town of Casitas. They announced that their intention was to force the federal government to resolve the crisis of the first blockade.[35] This second barrier prevented traffic from reaching the towns of San Rafael, Martínez de la Torre, and Teziutlán to the north, while to the south the Laguna Verde plant was completely cut off. Plant technicians and officials were trapped, including Dr. Rafael Fernández de la Garza, general director of the Laguna Verde project.[36] The truckers explained that the original blockade was costing them a great deal of money. They were so angered by the situation that José Javier Palafox Pozos, the delegate of their union (the Sindicato Nacional de Trabajadores de Transporte [SNTT]), declared that the union supported the forcible removal of the blockade.[37] Palafox Pozos explained that union members respected the opinion of the activists, "but it [the union] will not permit the blockade to further affect the interests and obligations of work."[38]

Seventy-two hours after the blockade began, the activists remained firm in their demand for a meeting with President de la Madrid. José Carrasco Flores, president of the Grupo Quetzalcoatl of Veracruz, announced that his group would not lift the blockade until the protesters received a favorable response to their request, a meeting with the president.[39]

As the blockade continued, rumors began to circulate about the possibility of military intervention,[40] and indeed the government had already begun to increase the military presence in the area. In a conciliatory move, the governor of Veracruz offered the protesters a meeting with either the head of the Secretaría de Desarrollo Urbano y Ecología (SEDUE) or the head of Gobernación (the Interior Ministry).[41] But protesters Marta Lilia Aguilar and Juan Marín remained firm: "We want to speak directly with Licenciado de la Madrid."[42] The president refused.

After four long days, the blockade was broken up by the military, but accounts of the forced lifting of the blockade vary. The governor and military officials depicted the end of the blockade as peaceful and orderly. The governor maintained that the dismantling of the blockade was necessary because five billion pesos were lost due to the obstruction of traffic.[43] The protesters were dislodged by the state's security forces, the army, the navy, and the Policía Federal de Caminos (Federal Highway Police). Governor Gutiérrez Barrios declared that although he had been compelled to call out the armed forces, the military had ended the blockade peacefully.[44] "I made them see that their action, besides being illegal, provoked discontent in the population, and there was a possibility of a confrontation between the truckers and the protesters," he said. "Even during the early hours today I myself offered to accompany a group of them to a meeting with a head of a ministry. But with the negative response from the leaders I had to make the decision of proceding with the break up of the blockade because what happens in the state is the responsibility of the governor."[45]

Antinuclear leaders narrated a different version of the events that occurred early on the morning of June 23, 1988. Genaro Guevara of the Grupo Antinuclear of Xalapa, José Arias Chávez of the Pacto de Grupos Ecologistas, Alfonso Ciprés Villarreal of the Movimiento Ecologista Mexicano, José González Torres of the Partido Verde Mexicano (Mexican Green Party), and Alejandro Calvillo of the Campaign Against Laguna Verde denied reports that the groups had accepted a meeting with Manuel Camacho Solís of SEDUE.[46] They also described the military's actions as "authoritarian and violent":[47] when "the agents arrived . . . they scattered what there was in the camp in a violent fashion; they pushed several vehicles into ditches and threw a pickup truck into the river."[48] The protesters had previously agreed that they themselves would not use violence and instead started singing as the armed forces began to dislodge them from the highway. The police arrested no one but impounded twelve vehicles belonging to *ejidatarios*[49] and cattlemen, threatening them with possible court proceedings and jail terms. Despite the violence and the threats, the antinuclear leaders believed their action was a victory for popular participation and a defeat for the government.

After the removal of the blockade, a minor schism developed among the various groups. The Madres Veracruzanas had agreed to a meeting with Manuel Camacho Solís of SEDUE despite the fact that many of the other groups objected. The Madres Veracruzanas, however, stated that the movement should be flexible and pursue as much dialogue with the various levels of government as possible. They also reasoned that Camacho Solís was close to the PRI's presidential candidate, Carlos Salinas. Most of the other groups, however, were frustrated about not having secured a meeting with the president and saw further meetings with heads of ministries as unproductive and unacceptable. Despite the criticism, the Madres attended the meeting on June 27, 1988, and declared it a success. According to Patricia Ortega Pardo, "The dialogue with Manuel Camacho Solís was fruitful; there were no baroque or byzantine discussions and for the first time we obtained two concrete responses to our demands: the promise of the meeting with Miguel de la Madrid and the assurance that SEDUE will provide information to the environmental and antinuclear groups. It will be information from [Camacho Solis's] ministry; he will offer us his reports on the environmental impact that Laguna Verde would have if it were to operate."[50]

All of the activists then reflected on their experiences with the blockade and the military. Movement participants were offended by official news reports indicating that the military personnel who broke apart the blockade were unarmed. They described armed military personnel threatening them with weapons, as well as destroying their belongings and damaging their vehicles. Nevertheless, the experience reinforced their pacifist beliefs: they intended to take the moral high road. Fernando Jácome of the Coatepec antinuclear group declared that the activists would retaliate against the government by voting for presidential candidates who would cooperate with the movement, such as Cuauhtémoc Cárdenas of the Frente Democrático Nacional (National Democratic Front [FDN]) or Heberto Castillo of the Socialist Party.[51] Jácome vowed that the struggle would continue until the population's antinuclear stance was heeded.[52]

Similarly, the Pacto de Grupos Ecologistas announced plans for new blockades of traffic in several locations to take place on July 6, election day.[53] Moreover, it, unlike the Madres Veracruzanas, declared that its members would get involved in politics by encouraging voters to cast ballots for any party except the PRI.[54] José Arias Chávez, coordinator of the group, also indicated that his group planned other acts of civil disobedience to protest against Laguna Verde, such as marking up electric bills.

Despite the disruptions caused by the blockades, the organizations did not achieve their objective of meeting with the president. Movement leaders met numerous times with the governor of Veracruz and with the head of SEDUE, but neither of these individuals had the power to prevent the Laguna Verde plant from operating. Most of the antinuclear activists believed that only the president could stop the project and thus continued their antinunclear activities in order to secure such a meeting.

The PAN Holds a Referendum

Meanwhile, the Partido de Acción Nacional (PAN) was attempting to use the Laguna Verde problem for its own political gain. Leaders of the PAN decided to hold a referendum in the state of Veracruz; the vote was to be held on June 5, 1988, and would reveal the population's position on Laguna Verde. Although the antinuclear groups favored the referendum in the abstract, leaders of the movement decided against joining the PAN in holding the referendum. The Madres Veracruzanas remained firm in their refusal to join or work with political parties, and Pedro Lizárraga of the Movimiento Antinuclear y Ecologista of Veracruz argued that an honest referendum concerning Laguna Verde was virtually impossible.[55] He said that "we [the antinuclear movement] do not have the money, or the capacity, or the power to carry it out, and the state or the government does not have the moral quality to execute it."[56] "What is there behind Laguna Verde that does not allow discussion to take place?" he wondered aloud.[57]

Along with the referendum, the PAN announced its plan for Laguna Verde if Manuel Clouthier, the PAN's candidate for the presidency, were elected. Humberto Ramírez Robledo stated that his party would convert the nuclear plant into a gas-fired installation.[58] PAN leaders believed that a gas-powered electrical plant was a logical solution, given that the Cactus-Reynosa pipeline, which spans most of the country from north to south along the Gulf Coast, is located only eight hundred meters from Laguna Verde: "In that way the gas that PEMEX [the government-owned petroleum company] presently burns would be put to good use. The cost of the conversion, according to several studies, would be 250 million dollars, which in no way compares with the human lives that would be protected."[59]

The Madres Veracruzanas were not persuaded by political rhetoric, and it was at this point that they began the intense red ribbon campaign. They declared that the red ribbon would be their symbol of opposition to the Laguna Verde project, so in early June of 1988 they distributed nearly a thousand ribbons to places of business and private residences throughout the state, and thereafter always wore red ribbons to antinuclear gatherings to symbolize their continuing opposition to the plant.[60]

The PAN continued to court the antinuclear groups, asking them to sanction and participate in the referendum. All of the groups, except the small Grupo Arcoíris, rejected the offer, however. José Arias Chávez explained, "We are grateful to the PAN for its support and disposition, but we have already said that for all of the good intentions of the referendum, it does not help us because [it] has to be broad and democratic and should be run by a neutral entity."[61]

The referendum was held on Sunday, June 5, 1988, in the state of Veracruz. Voting booths opened at 8 A.M., and early in the morning the PAN announced that it hoped for a turnout of one hundred thousand people. It had divided the state of Veracruz into six sections: (1) the northern zone, including Papantla, Poza Rica, and Tuxpan; (2) the nuclear zone near the plant, including Cardel, Vega de Alatorre, and Palma Sola; (3) the Xalapa area; (4) the Córdoba and Orizaba area; (5) the port of Veracruz and Boca del Río; and (6) the Papaloapan River basin in the southern part of the state.[62]

An overwhelming percentage of those who voted were against Laguna Verde. In the port of Veracruz, approximately 18,383 people voted; of these, 15,995 were against the plant.[63] Overall, 74,504 people in the state turned out to vote, with 93.72 percent against the Laguna Verde project and 6.28 percent in favor.[64] The PAN announced that the referendum "erases the national myths that say that the people cannot organize themselves in Mexico."[65] For the PAN, the referendum was a success.

The newspaper *El Financiero,* however, pointed out that the PAN had not met its announced goal to lure one hundred thousand people to vote in the state.[66] Moreover, the newspaper questioned the results of the referendum: in the days immediately following the vote, contradictory voting results were announced. The newspaper argued that the results in Xalapa were especially disappointing because only 8,187 people had voted there.[67]

A regional newspaper, *El Diario del Istmo,* also argued that the PAN had failed. Before the vote, the opposition party had declared that a turnout of less than one hundred thousand would signal that the public was apathetic concerning the Laguna Verde issue. After the vote, the party claimed, "We don't doubt that opposition to the operation of the plant has multiplied, and although popular participation has declined since the nomination of the PRI candidate, nonconformity remains everywhere. [Yet] Veracruz is not Chihuahua and the PAN lost once again."[68]

Meanwhile, the PRI also took the opportunity to lambaste the PAN for its efforts. PRI officials accused the PAN of manipulating the Laguna Verde issue for political purposes in an effort to influence the July 6 presidential elections. Jorge Moreno Salinas, secretary general of PRI's State Directive Committee, said that this was "the worst farce that they have ever planned."[69]

Juan Lobeira Cabezas, PAN candidate for deputy for the eleventh district, nevertheless refuted the criticism. "The referendum on Laguna Verde that the party organized is a palpable example that the population of Veracruz does not want the plant to operate as a nuclear facility. We are giving the governor of the state statistical information, palpable proof that the citizenry of Veracruz does not want the plant."[70]

The Presidential Election

By May 1988 the presidential race was in full swing. Cuauhtémoc Cárdenas, the candidate representing the Frente Democrático Nacional, had initially been cautious about the Laguna Verde problem, but by the spring of 1988 he had begun to criticize the plant.[71] On May 8 he announced that he was completely against the Laguna Verde project. He also declared that politicians lobbying in favor of the plant were probably receiving bribes.[72] He made his announcement in Palma Sola, where hundreds of his followers blocked the coastal highway.[73]

On July 6, 1988, the presidential election was held. Early on, it appeared that Cárdenas might win, but eventually Carlos Salinas de Gortari was declared the winner. Most of the antinuclear activists perceived the PRI's victory as a major disappointment. Despite his loss, Cuauhtémoc Cárdenas, speaking a few weeks later, exhorted his followers to continue with the party's struggle and advised them not to join the PRI's government: "Administrative positions do not interest us. They will not convince us to form a part of a cabinet that is attempting to usurp power; we will keep the struggle alive in a thousand ways."[74]

The Laguna Verde Plant Is Suspended

Unexpectedly, on July 26, 1988, the antinuclear groups received some good news: Manuel Camacho Solís of SEDUE announced that in order to guarantee the safety of the people of Veracruz, the Laguna Verde reactor would not be loaded.[75] He also promised to inform the groups about any other information that his ministry might receive regarding the plant. The antinuclear groups were jubilant about this news; it seemed as if the movement had triumphed. Pro-nuclear officials, however, were outraged. Juan

Eibenschutz of the Comisión Federal de Electricidad (CFE) denounced the antinuclear movement and claimed that it was trying to prevent Mexico from developing economically and technologically.[76]

Yet, the divisions among the antinuclear groups were still apparent during the summer of 1988 despite attempts to minimize them. Fernando Jácome's comments to Manuel Camacho Solís about the situation reveal a rift within the movement as well as a climate of suspicion among the participants. Jácome, a leader of an antinuclear group in the town of Coatepec, informed Camacho Solís in a meeting on July 27 that he believed the movement was now being manipulated by political parties of various ideologies. He argued, moreover, that foreigners were infiltrating the ranks of the various antinuclear groups.[77] These foreigners—Central Americans—allegedly were attempting to destabilize the country. Jácome specifically named the Nicaraguan Jorge Espinoza, saying that he was one of the people who had infiltrated the movement to create confusion by spreading beliefs that had nothing to do with environmentalism. Jácome blamed these foreigners for the idea of the blockade as a tactic: "The taking of highways was something that was not thought through. There are many ways to demand the total closing of Laguna Verde without the need to involve third parties."[78]

The Madres Veracruzanas tried to bridge the gap between the government and the movement. In an open letter, the Madres used a conciliatory tone: "Let us hope that from now on people and government will go hand in hand working for solutions to problems, leaving aside personal or group interests, for the good of all Mexicans. Madres Veracruzanas."[79]

The other antinuclear groups were now cautiously optimistic rather than jubilant as they began to scrutinize the government's announcements. The activists noted the cautious proviso that the reactor would not be loaded "for the moment."[80] Both antinuclear organizations and members of opposition parties regarded the announcement as a step in the right direction, though not a clear victory. At a political rally in Xalapa, Cuauhtémoc Cárdenas announced that the suspension was a positive but insufficient step.[81] Manuel Clouthier of the PAN argued that the suspension of the loading of the reactor was a purely political act.[82]

Many of the plant's opponents believed that the government had decided not to load the reactor because it was afraid of offending voters before the upcoming municipal elections. Because of this belief, José Arias

Chávez reassured the press that the struggle would not cease until the plant was closed permanently. Juan Marín went further, explaining that he feared the reactor would be loaded immediately after the municipal elections.[83] He announced that another national meeting of antinuclear activists—an informational meeting to evaluate the work that had been done and what had been achieved—would be held in Palma Sola in order to ensure that their voices would be heard.[84]

Not surprisingly, CFE officials and labor leaders had different reactions to the government's announcements. CFE's Juan Eibenschutz warned that nuclear energy was useful and necessary because the continuing use of coal and petroleum as energy sources could bring on disastrous consequences for the environment.[85] He also reiterated his position regarding the plant—that Laguna Verde was safe and that nuclear energy was crucial to Mexico's progress, contrary to the opinion of the "small group of ecologists who are opposed and who see only their own interests."[86] To back up his position, he explained that the European Community had saved the equivalent of 130 million barrels of petroleum through the utilization of nuclear energy.[87] A third official, José Luis Alcudia, the undersecretary at the Secretaría de Minas e Industria Paraestatal (Ministry of Mines and Industry [SEMIP]), did not mince words: the federal government would definitely put the plant into operation.[88] He revealed that the plant was practically completed and met international quality control standards.[89] Also, he argued, Laguna Verde should be permitted to operate because electricity should be produced by diverse methods; Mexico should not be forced to rely on hydrocarbons alone.[90]

Carlos Smith, regional coordinator of the labor union SUTERM, also joined the debate, arguing that human and other resources were being wasted by the suspension of the loading of the nuclear reactor.[91] He declared that twenty-five thousand workers were unemployed and that $180 million a day was being lost due to the suspension. He also indicated that highly trained workers had been leaving the plant: these workers were trained abroad with Mexican tax money, and now they had to leave the country to look for work.[92] Like others in favor of nuclear energy, he blamed this situation on the antinuclear movement and argued that movement activists were frightening the local population of cattlemen and peasants in order to get them to sell their lands at low prices.[93]

The antinuclear groups, skeptical that the government would actually cancel the Laguna Verde project, decided to proceed with their protest activities. At a meeting on the Universidad Veracruzana campus on August 7, 1988, they decided on their plan of action for the near future: they would organize voluntary blackouts in homes and businesses every day between 8 P.M. and 8:15 P.M. in Xalapa and throughout the state to protest the Laguna Verde project. They also agreed to continue blockading roads and highways not only in Veracruz but throughout the country.[94]

In addition, the activists discussed the issue of political affiliation. Although overall the antinuclear groups maintained that they were not affiliated with any one particular party, they nevertheless agreed they would have to support political candidates who declared themselves to be opposed to the nuclear plant and nuclear technology in Mexico.[95] Efraín Romero, Pedro Lizárraga, Roberto Helier, and José Arias Chávez announced that in the municipal election of October 8, 1988, they would not vote for any candidate who openly supported Laguna Verde. They felt morally and ethically bound to support opposition parties, such as the PAN, because these parties were firmly opposed to the plant. To protest the electoral fraud perpetrated by the PRI in the July 6 presidential election, they would participate in political rallies held by the opposition parties.[96]

José Arias Chávez also challenged the statements made by Carlos Smith of SUTERM regarding the amount of money lost by the suspension. He maintained that only $180,000 a day was being lost, as opposed to the $180 million declared by Smith.[97] Roberto Helier also criticized the CFE for going over its original budget of $200 million by $600 million.[98]

The antinuclear groups not only spent time debating the merits of nuclear technology and protesting against Laguna Verde but also served as watchdogs for any problems at the plant. Through information surreptitiously provided by a few anonymous technicians working at the plant, the groups acted quickly to announce problems at Laguna Verde. On August 8, 1988, Roberto Helier stated that the plant had suffered a small mishap in its cooling area.[99] When one of the four pumps in the system was tested, it failed, which led to a decrease in the water level of the basin (*dársena*) located next to the Gulf waters. According to Helier, this situation would have been extremely dangerous had the plant been operating.[100]

Marco Antonio Martínez Negrete, in agreement with Helier, spoke of the accident that occurred on March 9, 1988, at a nuclear plant in La

Salle, Illinois. He saw similarities between the accidents at La Salle and at Laguna Verde, saying it "destroys many of the arguments of the pronuclear forces."[101] The Illinois Boiling Water 5 Mark II reactor—also designed and built by General Electric—was very similar to that at Laguna Verde. At La Salle, the circulation pumps in the cooling system experienced problems. The antinuclear activists warned the population that the Laguna Verde plant's design was inherently unstable and flawed.[102] In fact, they knew that GE engineers themselves had declared Laguna Verde's reactor model to have design flaws.

Only one month after the suspension of the loading of the Unit I reactor, the antinuclear groups became increasingly suspicious that the plant might go on-line relatively soon.[103] They said that high- and middle-level technicians at the plant had indicated that the government still planned to operate the plant. Eugenio Martínez, former head of Radiological Security at the plant, declared that the plant should not be put into operation, given that at every phase of its construction two or three serious problems had emerged.[104] Roberto Helier agreed but warned that movement activists had been informed by technicians at the plant that despite the problems, Laguna Verde would begin operating soon.[105]

Similarly, the social scientist Gustavo Esteva suspected that the Laguna Verde plant would probably be put into operation, despite SEDUE's announcement to the contrary.[106] He exhorted the activists to stay mobilized until the government agreed to cancel the nuclear project and maintained that technology should be under popular control: "This is a part of democratic government. We want to free ourselves of the dictatorship of [technical] experts."[107]

Marco Antonio Martínez Negrete also reflected on the government's motive for suspending the loading of the reactor, suggesting with pessimism that the reason was entirely technical: perhaps there was a problem at Laguna Verde similar to the problem at the La Salle, Illinois, plant. "Clearly, a government worried about the health and well-being of the country would have cancelled Laguna Verde solely because of the incident in the United States."[108]

Any optimism Martínez Negrete may have had stemmed purely from political factors: "The government realized that to initiate the loading and testing would mean that it would lose moral authority. The government was concerned about the impending municipal elections and public opinion."[109] The antinuclear groups hoped that their actions had provoked the permanent suspension of the loading of the reactor.

On September 1, 1988, President Miguel de la Madrid delivered his sixth and last Informe de Gobierno (Address to the Nation). The antinuclear activists then spent several days analyzing the speech, wondering why de la Madrid had not made reference to the Laguna Verde issue. There were two possible interpretations: (1) the plant could begin operation quite soon, still under the de la Madrid administration; or (2) the government had grown tired of the issue and would not approve the operation of the plant. José Arias Chávez of the Pacto de Grupos Ecologistas warned that "the antinuclear activists and environmentalists should not believe either one blindly, but instead should intensify the struggle and not let down their guard."[110] He had confidential information that opposition members in the Chamber of Deputies were considering appealing to the president about the Laguna Verde issue. The Madres Veracruzanas suspected that the presidential address and in particular the president's silence on the issue indicated that the decision to operate the plant despite opposition had been made behind closed doors, María del Carmen Calvo Barquín of the Veracruz chapter of the Madres declared.[111] Arias Chávez also stated that Feliciano Béjar and Ofelia Medina of the Grupo de los Cien would meet with an official from the president's office to ask for a meeting with the new head of SEDUE, Gabino Fraga Mouret.[112]

Two days later, antinuclear activists were again somewhat optimistic when James Puryear, an American engineer, consulted with the Mexican government about the possibility of converting Laguna Verde from a nuclear to a conventional plant.[113] Puryear assured the press that the Laguna Verde installations were obsolete, in particular the reactor with its now twenty-year-old design. Movement participants believed that Puryear's visit to Mexico was a positive sign because government officials appeared to be considering alternatives to nuclear technology.

In early September 1988, Horacio Lombardo Pérez Salazar, an adviser to President de la Madrid, hinted that Laguna Verde might be given permission to operate in the near future, maybe even within the next three months, before Carlos Salinas was scheduled to take office.[114] Lombardo warned, however, that the plant could function only after a careful coordination of the positions held by federal and state governments and the population as a whole. If these three constituencies could not come to some mutual agreement, the plant's operation would be delayed once again: the president had always maintained that this type of project should be carried out only with the approval of the population.[115]

The Madres Veracruzanas responded angrily to this announcement. Rebeka Dyer, of the Xalapa chapter, retorted that de la Madrid's adviser obviously did not understand the position of the people of the state of Veracruz:[116] "The antinuclear groups from throughout Mexico have the word of the president, Miguel de la Madrid, that Laguna Verde shall not be put into operation for the moment and that before making the decision of loading the reactor, the people shall be consulted. Miguel de la Madrid's word was passed on to us by Manuel Camacho Solís when he was head of SEDUE, and it was backed up by Governor Fernando Gutiérrez Barrios."[117]

Meanwhile, CFE and SEMIP officials began telling the press that they were sure the Laguna Verde plant was ready to operate; all of its technical problems had been solved, but the political problem remained. Fernando Hiriart, head of SEMIP, announced that he believed the plant had only a 50 percent chance of being put into operation under the de la Madrid administration because of political problems.[118] He stated that the government was beset by many problems and was probably hesitant to take on the one posed by Laguna Verde.

Thus, in an apparent move to address this public relations problem, the CFE invited antinuclear groups to visit the nuclear installations. CFE officials reassured the antinuclear activists that the plant was safe and did not represent any type of threat to the population.[119] They also announced that the plant had a laboratory with equipment—purportedly among the most advanced in the world and purchased at a cost of six billion pesos— to measure radioactivity in the environment. Thirty technicians would constantly be monitoring the situation and were to advise the CFE heads of any problems in the environment. Operations at Laguna Verde would be modified to address these problems if detected. The CFE, also aware that the antinuclear groups were extremely concerned about radioactive waste disposal, said that the plant installations were capable of storing nine hundred tons of radioactive waste in the "pools" *(albercas)* adjacent to the reactor for fifty years. These wastes would then be buried once they had lost their high toxic content "in totality."[120]

The antinuclear groups were not convinced, however, and Homero

Aridjis of the Grupo de los Cien denounced the CFE and SEMIP, saying that these institutions were actively involved in a "campaign of disinformation"[121] in order to ensure that Laguna Verde would be put into operation. He also denounced the censorship suffered by the anti-nuclear movement—specifically by the Grupo de los Cien. First, a press release sent by the group to various newspapers was never published. This press release declared that the Grupo de los Cien did not believe that Laguna Verde should be allowed to operate without a referendum in the state of Veracruz and a public debate in the Chamber of Deputies. Also, "the government should learn the lessons of the July elections and not make antidemocratic decisions that go against the will of the people."[122] Aridjis found the censorship especially problematic because the government had congratulated itself on its democratic opening—that is, what it considered its move toward greater democracy: "But the cases of censorship aimed at individuals, institutions, parties, publications, and religious creeds have gotten worse, and it is worrying when one examines the long list of journalists assassinated in different parts of the country during this administration."[123]

Rumors About the Reactor

By late September 1988 the antinuclear groups had received confidential information that the reactor would be loaded on October 15. The Madres Veracruzanas contacted the press, urging the population to make known its disapproval of the Laguna Verde project. They urged citizens to make their views known immediately because rumors were circulating about the imminent loading of the reactor.[124]

A candidate for the municipal presidency of the port of Veracruz backed up the Madres and revealed the government's plan to build public support for Laguna Verde. María del Carmen Cover Hernández of the Partido Auténtico de la Revolución Mexicana (PARM) announced that she was against Laguna Verde and accused the government of "provoking the citizenry."[125] She attributed the suspension of the loading of the reactor to public outcry against the project. However, the government was still clearly determined to operate the plant. "They can say what they want," she pointed out, "they can argue about the important advances, but they have not been able to demonstrate that Laguna Verde, like any nuclear plant in the world, is safe."[126]

Rumors about the imminent loading of the reactor continued to circulate, prompting federal government officials from SEDUE and the CFE to announce at a press conference that the rumors were untrue.[127] Mariano Luis Fuentes, private secretary to the head of SEDUE, Gabino Fraga Mouret, said that although it was true that Ofelia Medina and Feliciano Béjar of the Grupo de los Cien had met with Fraga Mouret, there was a misunderstanding about the loading of the reactor; the exact date for loading had not been decided.[128] Skeptical, Jesús Rodal of the Cattlemen's Association of Palma Sola countered that the members of the Grupo de los Cien had received a report that the reactor would be loaded on October 14 or 15.[129]

The Grupo de los Cien did not believe SEDUE's announcement and expressed alarm about the impending operation of the plant. According to Homero Aridjis, loading the reactor would be tantamount to "a coup against the people."[130] He added that "the government has not learned its democratic lesson from July 6 and continues employing antidemocracy."[131]

The leaders in the antinuclear movement immediately headed for the Senate to prevent the executive branch of the federal government from making a unilateral decision, a common occurrence in Mexico. They informed the senators that the plant should not be allowed to operate until the thirty-five thousand *amparos* (injunctions) filed in Veracruz were decided.[132]

The debate became heated, and pro–Laguna Verde officials began to lose their patience. Leonardo Rodríguez Alcaine, the head of the electrical workers' union, SUTERM, lashed out angrily at the antinuclear movement. He maintained that the people of Mexico as a whole were not opposed to nuclear energy. Those responsible for opposition to the plant were in fact a very small group of "so-called ecologists"[133] who were using the Laguna Verde issue as a smoke screen; their real motive was to form a party, the Green Party, in order to pursue their political ambitions. He denied that the plant was obsolete or that it had technical problems and dismissed those who criticized the plant: "Whoever says that [the plant is obsolete] is an idiot, and I am speaking very frankly—an idiot—because they have not seen the results that we have seen at Laguna Verde. We can assure [everyone], now with more reason than before because we have the general and international test results, that the plant is super safe, though for economic reasons we have not been able to operate it."[134]

The Comisión Nacional de Seguridad Nuclear y Salvaguardas (National Commission on Nuclear Safety), whose job it was to oversee the Laguna

Verde plant, also tried to assure the public that the plant was safe. Officials from the commission said that the plant was built solidly and was impervious to radioactive leaks, earthquakes, fires, or any other type of natural phenomenon.[135] "The plant has been built to the same standards applied to plants in the United States,"[136] according to Roberto Gonzalez, revealing his awareness of the sensitive issue of Mexico's position vis-à-vis the industrialized countries. "Laguna Verde will be the most complete model of quality control for electrical plants."[137]

Government officials appeared to be gearing up for the operation of the plant. On October 5, the commission confirmed that it had granted permission to the CFE to load the first Laguna Verde reactor with uranium 235 and 238.[138] The electrical workers' union also gave signals that Laguna Verde would begin operation soon: "We would not take risks with our lives; if we knew that the plant did not meet the international and national requirements for safety, we would already have said so. We are true professionals and much more honest than those groups who call themselves 'ecologists.'"[139]

As the loading of the first reactor seemed more and more imminent, the national newspaper *El Financiero* deplored the fact that the emergency evacuation plan for the plant, the PERE, had still not been improved. According to Avelino Hernández Vélez of *El Financiero,* the PERE was still woefully inadequate, despite its approval by the commission.[140] He blamed the faulty evacuation procedure for raising the local population's ire against the plant. Doctors working in clinics and hospitals near the plant had said they would be the first to abandon the area in the case of an accident, although their health centers were supposed to be responsible for treating the injured. In addition, Hernández reminded readers that roads near Laguna Verde that were supposed to serve as evacuation routes were often not passable year-round, and many were not functional at all during the rainy season from June to October. Finally, Hernández argued that some of the recommendations made by the PERE were "absurd" and had offended the local population—for example, residents were to shut all windows and doors to prevent contaminated air from entering even though many of the houses in the area were of poor quality and "were not designed to prevent breezes from entering."[141] The local residents were acutely aware of all these problems.

In a last-ditch effort to dissuade the government from loading the reactor, Homero Aridjis of the Grupo de los Cien also spoke out. He said the movement believed that a referendum, not a bureaucratic decision made by the CFE, should determine the plant's fate.[142] He argued that the CFE was attempting to take advantage of the "power vacuum during [the transition period] of one president leaving and another entering" the political arena.[143] The plant's operation at this time would be tantamount to a defeat of the people because the government had not consulted with civil society.[144] Moreover, given that the plant had experienced numerous construction delays and that many companies had been involved, Aridjis called it "a patched up monster" that was a menace to society.[145] Alfonso Ciprés Villarreal, president of the Movimiento Ecologista Mexicano, added that he was skeptical about the government's position regarding Laguna Verde. He believed that the suspension of the loading of the reactor had been a cynical political move designed to elicit momentary "applause"[146] during the campaigns before the municipal elections.

As tensions rose about the fate of the Laguna Verde plant, the federal government warned antinuclear activists about the seriousness of engaging in illegal protest activities. José Luis Sáenz Escalera, a government spokesman, specifically warned them about the illegality of certain protest tactics, such as blocking highways and taking over radio stations and government buildings.[147] No matter what the circumstances were, blockading highways was a serious crime according to the federal penal code. According to Sáenz Escalera, "rebellious groups"[148] were finding it easy to use these tactics as an instrument of pressure in order to protect "their supposed rights."[149] He believed that a hard line should be taken in the future against "agitators" who engaged in these activities: "We have indicated that there should be a firm hand against those who execute these criminal acts."[150] Clearly, the government had erred in its response to the last Palma Sola blockade; the leaders should have been arrested in order to prevent these activities from occurring in the future. Through its spokesman, it appeared that the government was signaling the antinuclear movement about the danger of pursuing its favorite protest tactics.

The Veracruz Municipal Elections

The Madres Veracruzanas continued their work as the municipal elections approached. In a paid advertisement in the *Diario de Xalapa*, they reminded the citizenry of the importance of the elections. Because they always maintained that they were "above politics" and not affiliated with any political party, they did not endorse any candidates. Rather, they said, "Pick those whose experience, aptitude for conciliation, and vocation for service is demonstrated, and who furthermore have pronounced themselves to be against the Laguna Verde project. . . . Only by voting will you be participating."[151]

Local and national newspapers acknowledged that the municipal elections would be tense. The PRI had a long history throughout the country of fraudulently ensuring the victory of its own candidates in cases where opposition candidates may have won. Thus, given the fact that many citizens were still disturbed by their perception of electoral fraud in the July 6 presidential election, observers worried that the elections in the 204 *municipios* (roughly like U.S. counties) of Veracruz would not be smooth. The Laguna Verde issue only made matters worse because many of the opposition candidates had declared themselves to be against the plant.[152]

An announcement made by the antinuclear coalition on October 3, 1988, exacerbated the tensions. José Arias Chávez and several other leaders of antinuclear groups throughout Veracruz and from the states of Jalisco and Tabasco demanded that all candidates in the municipal elections declare their positions regarding the Laguna Verde issue.[153] The activists had also agreed unanimously that the only solution to the alleged electoral fraud in the presidential election of July 6, 1988, would be to ask for an annulment of those elections, which had resulted in the victory of the PRI candidate, Carlos Salinas de Gortari. To back up these statements, the antinuclear groups, including the Movimiento Antinuclear del Estado de Veracruz (MAEV) and the Madres Veracruzanas, held a protest march on October 3, 1988, in which they declared themselves to be against Laguna Verde, against electoral fraud, and against the injustice that had occurred during the massacre of students in Tlatelolco on October 2, 1968—one of the most notorious political incidents in Mexican history.[154]

The antinuclear activists were subsequently encouraged when opposition leader Porfirio Muñoz Ledo proposed that the Mexican Senate invite the public to participate in the nuclear debate within that body.[155] He argued that Article 58 of the internal rules of the Chamber of Deputies called for open meetings of the Comisiones Unidas de Energéticos (Commissions on Energy Sources). He hoped that if the meetings were open to the public, the Senate would not approve the operation of Laguna Verde. In the meantime, the results of the municipal elections were slowly being released: many of the antinuclear activists were disappointed not only in the results themselves, but also in what happened after the results were in, and they declared that some of the candidates had actually deceived them.[156] Genaro Guevara of the Grupo Antinuclear y Ecologista of Xalapa and Victor Meza of the Pacto de Grupos Ecologistas cited the case of one successful candidate, Héctor Ortiz of the Partido Popular Socialista (PPS). Ortiz had run as an anti–Laguna Verde candidate, then switched his position after the election. The activists lamented the fact that certain politicians had taken advantage of the Laguna Verde issue to help themselves get elected. Genaro Guevara stated, "For this reason we believe it is necessary for parties and deputies to define themselves and explain when they are speaking for themselves and when they are representing their parties."[157]

The results of the municipal elections were indeed disappointing for the antinuclear movement. Elections were held in 198 municipios throughout the state of Veracruz, and of these the opposition won in only 7 municipios.[158] Voter turnouts were also disappointing—only 31 percent of the eligible voters actually went to the polls. According to the Comisión Estatal Electoral (State Electoral Commission), of the seven defeats for the PRI, three were meted out by the Mexican Socialist Party (PMS), two by the PPS, and only two by the Partido de Frente Cardenista de Reconstrucción Nacional (PFCRN).[159] Two factors could explain these results. First, electoral fraud has been a fact of life in Mexico throughout this century, and these municipal election results probably reflect that factor. In addition, the high rate of voter absenteeism also revealed the population's disillusionment with the political system. During the late 1980s, local intellectuals and opposition leaders often pointed to absenteeism as a symptom of the population's lack of confidence in the political system. It seems that although many citizens of Veracruz were worried about the plant, few believed that voting in elections would solve the problem.

Villagers Fear Evacuation Plan

While antinuclear groups debated the government and the CFE over the fate of the plant, the people of the village of Palma Sola were truly racked by fear. They lived less than two kilometers from the plant, and the entire debate was making them so fearful that local doctors were noting a grow-ing number of people with gastrointestinal problems caused by stress and panic.[160] Parents did not allow their children to attend classes because the evacuation procedure pamphlet, the PERE, had scared them too much with its instructions that in case of a nuclear accident, they should not pick up their children at school because evacuation officials paid by the govern-ment would take them directly to shelters. The people of Palma Sola were convinced that there would be an accident if the plant were to operate, and they were panic-stricken over the thought of being separated from their children under those circumstances.[161]

The Madres Veracruzanas had worked closely with the cattlemen of Palma Sola and had a deep sympathy for the villagers' problems. Rebeka Dyer attributed the villagers' sense of panic to several factors.[162] First, there was the problem of false information: the people of Palma Sola had been fed much disinformation, especially in the PERE. Second, ministry heads responsible for Laguna Verde had not treated the PERE seriously enough. Finally, the village had been inundated with phone calls from throughout the country and from abroad, asking for news about the plant. These factors, as well as the increased presence of the military, had caused the villagers to go into a panic. Pedro Lizárraga of the Pacto de Grupos Ecologistas warned that the village population was so worried about the plant that residents were considering fleeing the area if the plant were to begin operation.[163]

Despite their fears, the residents of Palma Sola continued speaking to the press. Parents of schoolchildren declared that they would keep their children out of school as a protest against Laguna Verde. The director of Palma Sola's school, Alejandro Romero Castillo, confirmed that absen-teeism was high and denounced the evacuation plan: "In case of a nuclear accident during class hours, the teachers cannot make themselves respon-sible for the students; neither can we guarantee their safety."[164] In his statement Romero referred specifically to the guidelines of the PERE: ac-cording to the plan, school officials would be responsible for guiding the students to the proper shelters. Townspeople also decried the lack of shel-ters and hospitals in the area: the PERE's guidelines were based on the existence of a larger number of both types of facilities. Jesús Darío Rodal Morales, head of the local Cattlemen's Association, also denounced the presence of the military in the area. "This just demonstrates the dictatorial state of the government."[165] A town doctor, Alfonso Romero Perea, also warned that because the area was rural and lacked good roads, it would be necessary for townspeople to pass right next to the nuclear reactor on their way to shelters.[166]

Although antinuclear activists expected the reactor to be loaded soon, officials from the president's office denied that the plant would be operating in the next few days. However, in early October, when José Luis Alcudia, head of SEMIP, testified before the Chamber of Deputies that the installations were practically ready to begin operation, Porfirio Muñoz Ledo of the PFCRN and Feliciano Béjar and Ofelia Medina of the Grupo de los Cien left the building in a huff, convinced that the federal government had secretly told CFE officials to continue with their preparations in anticipation of the loading of the reactor in the near future.[167] Although no government official had announced it, the loading of the reactor seemed slated for October 11 and 12. Meanwhile, the plant remained shut down, seemingly to resolve last-minute details before the loading of uranium 235 and 238.

Nevertheless, antinuclear groups continued with their attempts to forestall the operation of the plant. The Grupo de los Cien, along with the Partido Verde Mexicano, continued to tell the press that proceeding with the Laguna Verde project would show the continued authoritarianism of the federal government. Homero Aridjis of the Grupo de los Cien said that the day the reactor would be loaded would be "a black day for democracy in Mexico"[168] and that "the imposition could lead us to a climate of generalized violence."[169] He believed that pro-nuclear forces from SEMIP, the CFE, and SUTERM were exerting strong pressure on President Miguel de la Madrid against the will of the Mexican people. He also wondered where the so-called "democratic opening" of the de la Madrid administration had gone.[170] Jorge González Torres, leader of the Partido Verde Mexicano, agreed that the Laguna Verde case had not been handled democratically: "I affirm that it is not an isolated case because in the general panorama of the country the popular will has been attacked on several occasions. The most recent case was seen in the electoral process of the sixth of July."[171]

Despite the volley of arguments from both sides, work at the reactor continued. The plant was made ready so that the uranium could be loaded at a moment's notice. Ninety tons of uranium were placed near the reactor, and the installations continued to be tested.[172] To make matters worse for Palma Sola, the municipio offices announced that just before the reactor was loaded, eight thousand residents of Palma Sola would be evacuated.

This announcement led once again to a large protest on the coastal highway: at nearly a moment's notice, hundreds of antinuclear activists appeared to protest.[173] They demanded that the plant be converted to a conventional coal- or gas-fired installation. In response to the activists' mobilization, the Mexican army arrived with hoses and other fire-fighting equipment and lined up next to the protesters.

While antinuclear groups continued to mobilize and to speak to the press, the executive branch of the government and other relevant ministries remained silent on the issue. On October 11, 1988, Gabino Fraga Mouret, head of SEMIP, said only that nothing concrete had been decided about the plant's fate and that the federal government would make the final decision.[174] The three ministries involved—SEMIP, the CFE, and SEDUE—would merely present reports to the Comisiones Unidas de Enérgeticos from the Chambers of Deputies and Senators, but ultimately the politicians, not the technicians, would make the decisions.

On October 12, 1988, hundreds of antinuclear activists from Veracruz and Mexico City attended the debate over Laguna Verde at the Chamber of Deputies.[175] They spoke to various deputies, trying to convince them that Laguna Verde should never operate. Nevertheless, the Chamber approved a resolution supporting Laguna Verde. It also recommended that before the uranium was loaded, the Comisión Nacional de Seguridad Nuclear y Salvaguardas should reassure the deputies that the plant did not pose a risk to the population or to the environment.[176] It proposed the formation of a multiparty group to visit the plant, report on its condition, and organize meetings between the legislators and plant officials. Opposition members simply could not outvote the PRI delegates, but they nevertheless expressed their reservations about the plant and their dismay at the presence of the military in the Palma Sola area.[177]

The antinuclear groups continued their activities to prevent the plant from operating. The cattlemen from Palma Sola focused their efforts on the evacuation procedure, the PERE. In order to better evalutate the PERE, they decided to travel through the region visiting the shelters and clinics mentioned in the document. They found that not only were the shelters inadequate to protect people in case of a nuclear accident, but medical personnel—doctors and nurses at the clinics—were unaware of the appropriate emergency procedures.[178] Jesús Darío Rodal Morales, the cattlemen's representative, announced that the eighty-one cattlemen's associations in the state, totaling twenty-five thousand members, had already informed the federal and state authorities of their findings, and they demanded a referendum to decide the fate of the plant.[179]

Hoping that the decision about Laguna Verde would be made by the Chamber of Deputies, not by the president alone (as is often the case in Mexico), participants in the antinuclear movement sent that body a message to explain the reasons for their opposition to the plant. In the letter they explained the scientific basis for their opposition, and all of the antinuclear groups, along with other popular organizations, asked the Chamber to consider the Laguna Verde case more carefully.[180] The letter included seven basic points criticizing the project and the logic of CFE officials. Along with the usual arguments about the inherent risks of nuclear energy and the great expense of the project, the groups also raised new points. First, in response to the CFE's position that the project was operating within the boundaries of the law, the antinuclear groups countered that laws and rights were indeed being violated: fundamental rights of Mexican citizens such as "Article 4 of the Constitution [right to health], Article 14 [right to life], Article 16 [guarantee of legality], Article 17 [right to life], Article 27 [right of peasants to work their lands], and Article 39 [right to democracy]."[181] The groups stated that "according to Article 39 of our constitution, sovereignty resides with the people and, like any constitutional right, does not lapse, so the people have the ability to decide about Laguna Verde. The majority of the people are against Laguna Verde as opinion polls, referenda, signatures, amparos, blackouts, marches, protests, etc., have demonstrated. But if doubt persists, a national referendum would be indicated to resolve it."[182]

The letter's last line was, "In these critical moments, as in so many others in the history of the fatherland, the sovereignty of the people is represented by all of you; we ask that you exercise it in favor of your constituents, who demand the immediate suspension of the nuclear project."[183] During the early years of the movement, the activists had made mostly scientific arguments against the plant, but by 1988 they were engaging in political analysis. Although they still referred to scientific arguments, they now emphasized much more the political nature of the Laguna Verde issue.

On October 13, 1988, the Chamber of Senators approved the operation of the plant. The opposition did not have sufficient votes because the majority of the Senate were PRI members. Members of the Grupo de los Cien were present, lobbying the senators to reject the plant. Upon hearing the announcement, Ofelia Medina, a leading member of the Grupo de los Cien, yelled out that the senators were "irresponsible and undignified representatives of the people."[184] The situation quickly deteriorated, with pro- and antinuclear forces screaming at each other. The environmentalists cried out, "senadores asesinos" and "criminales." Leonardo Rodríguez Alcaine of the electrical workers' union yelled, "Saquen a esa vieja loca" ("Remove that crazy old woman"), referring to Ofelia Medina.[185] This comment would upset movement members for a long time. As Medina left the Chamber, she declared in tears that this was "the prelude to a holocaust"[186] but that the Grupo de los Cien would continue in the struggle to close the plant. Porfirio Muñoz Ledo of the Frente Cardenista pointed out that the senators had not complied with the wishes of the public that all debates be made public. Additionally, he insisted that the technicians had still not reassured the public about safety concerns.[187]

With every announcement about the impending operation of the plant, the people of Palma Sola grew more fearful. The military was patrolling the area around the clock, causing more fear and panic. Ambrosio Cortés, a resident of the area, said that approximately one hundred soldiers patrolled the beaches, towns, and ranches near the plant, and they made the villagers anxious and fearful because they were armed.[188] Jaime Castillo, of the local cattlemen's group, explained that the price of land had plummeted in the area because many people were selling their lands quickly to get out. Residents were also trying to move to areas as far away as possible because of the tyrannical behavior of the authorities.[189] Castillo expressed fear that once the plant had begun to function, wholesalers and customers might reject the beef and milk produced in the area, which would lead to an economic collapse for the local agricultural economy.

CFE officials continued to attack the antinuclear and environmental movements. Carlos Granados, coordinator of the plant's office for Comunicación Social y Relaciones Públicas (Public Relations) described as absurd the antinuclear groups' proposal for converting Laguna Verde into a conventional plant run by gas or coal:[190] "It would be like installing a tractor motor in a race car; if we wanted to do it, it would be cheaper to abandon Laguna Verde and build a new plant. Further, it would not be cost-efficient because in order to generate the same amount of energy that Laguna Verde would produce, it would be necessary to use up all of the gas in the country; nothing would be left for gas stoves."[191] Granados also tried to taint the movement by insinuating that the antinuclear groups were engaging in protests only to gain the attention of national and international media.

On October 14, 1988, newspapers reported that the government's go-ahead for the plant seemed imminent because the federal government had sent an official bulletin to Fernando Gutiérrez Barrios, the governor of Veracruz, stating its "determination to initiate the process of the operation of the first unit of the nuclear plant at Laguna Verde."[192] In response to this announcement, the Madres Veracruzanas descended upon the Governor's Palace. They all politely asked the governor to defend them and Veracruz, but one of the Madres, Guillermina Jiménez, a local business owner, boldly went up to him and said, "Mr. Governor, let's speak *jarocho* to *jarocho*: pull up your pants, prevent Laguna Verde from operating, and you shall take your place in history."[193] This statement is well known throughout the region and still recalled fondly by movement activists. Despite the antinuclear movement's activities, however, it appeared that the federal government would operate the plant.

• • •

From early 1987 to October 1988, the antinuclear movement engaged in intense mobilization at both the grassroots and elite levels. The most visible manifestations of mobilization occurred during the three blockades of the coastal highway near Palma Sola and the innumerable protest marches in Xalapa, the port of Veracruz, and Mexico City. Though the last blockade was broken up by the military, the antinuclear groups were pleased that their cause had received newspaper coverage. The movement's activities were never on television news programs, however, because of heavy government censorship.

The government sent mixed messages to the activists during this period. At times during 1988, it seemed willing to abandon the Laguna Verde project, given the level of opposition to its operation. In the end, however, the executive branch (President Miguel de la Madrid) decided to proceed without even waiting for the Chamber of Deputies to finish its debate. Though the senators had approved the Laguna Verde project, the lower half of the Chamber—the deputies—had not yet considered the matter. The CFE remained steadfast in its support of the plant and even tried to get the public to believe that once the plant began to operate, the process was irreversible.

At the ideological level, the various actors attempted to frame the debate differently. The antinuclear groups simultaneously framed the debate with both technological *and* political arguments. On the one hand, they gathered information from within the plant (from anonymous sources) and from abroad that indicated the plant was plagued with numerous problems. The groups argued, therefore, that the plant should be shut down because its technology was flawed. On the other hand, they also framed the discussion in political/democratic terms. Over and over again during this period, movement leaders such as Homero Aridjis criticized the government for proceeding with the Laguna Verde project without consulting civil society, and thus questioned the government's own claims of a "democratic opening."

Pro-nuclear arguments were usually rooted in technological grounds. The electrical workers' union SUTERM and CFE officials downplayed the dangers of nuclear technology and extolled the virtues of the Laguna Verde plant. Carlos Smith of SUTERM repeatedly criticized the antinuclear groups for their "political" position, saying that they were simply attempting to satisfy their political and economic ambitions by forming a political party and by purchasing lands in the Laguna Verde area.

4 The Loading of the Reactor

On October 15, 1988, SEMIP finally announced that, indeed, President Miguel de la Madrid had given his permission to proceed with the loading of the first reactor. SEMIP officials said that the decision was based on the benefits that would accrue by pursuing the path of nuclear energy, including the conservation of oil and diesel fuels, mastery of a technology that sooner or later Mexico would have to employ anyway, and the prevention of the need to ration fuel in the future.[1] They also declared that the federal government had addressed the concerns of the antinuclear groups by providing "maximum guarantees"[2] that the plant was secure and would not harm local residents or ecosystems. Moreover, several bodies—including the International Atomic Energy Commission and the Comisión Nacional de Seguridad Nuclear y Salvaguardas—had recently approved the Laguna Verde plant for operation. According to SEMIP, the government had succeeded in convincing the residents that their security, health, and lives were not at risk.[3]

An editorial in the Mexico City daily *La Jornada* provided an interpretation of the timing of the announcement. The editorial argued that the sensitive decision of whether to operate Laguna Verde had been left to a lame-duck president, Miguel de la Madrid, to avoid burdening the Salinas administration early in the new *sexenio*.[4] The newspaper also warned that the federal government had missed a golden opportunity to exercise democratic procedures. Specifically, both the senators and the deputies had left their tasks unfinished. They had promised that debates in Congress would be public and that public opinion would be taken into account. *La Jornada* believed, however, that the democratic process had been shunted aside.

Predictably, the antinuclear groups were dismayed by the president's decision. The Grupo de los Cien announced that "once more it was demonstrated that democracy does not exist in Mexico, and the worst part is that there is talk of opening four more similar plants."[5] Determined not to give up, the groups made plans for further protests. All of the movement's participants—artists, housewives, cattlemen, peasants, intellectuals, and business owners—asked the population to join in on a protest rally in the main plaza of the port of Veracruz on the following day, October 16. Movement leaders declared that this rally would pressure the legislative body into denying permission for the operation of the plant.[6] They hoped that the legislature would be clear about its position. If it chose to grant the permission, the movement participants would then know for sure that the Congress was an ally of the president "and an enemy of the people."[7]

The Madres Veracruzanas clarified their own position, however. They argued that in no way would they urge the population to use violence or to confront the armed forces. They also no longer favored the tactic of blockading highways: "That would mean provoking the army, which for days has been patrolling the highways."[8] The Madres nevertheless protested the announcement by forming a human chain around the Governor's Palace.[9]

The antinuclear groups used pamphlets, radio announcements, and press conferences to mobilize the general population of Veracruz. Radio announcements urged the population to meet at 5 P.M. on October 16 at the Civil Registry of the port of Veracruz to decide democratically on tactics and strategy to use in the near future to prevent the operation of the plant.

As the discussion about Laguna Verde continued, yet another newspaper, the Xalapa daily *Política*, published an editorial expressing its interpretation of events. The editorial argued that although technicians and technocrats were attempting to portray the Laguna Verde problem as technical in nature, it was in reality explicitly political.[10] For the editors, the implications of the decision were political because the population was fearful and opposed to nuclear technology, yet the government had decided to proceed with the project in spite of the population's opposition. *Política* explained that the population was especially fearful because of Mexico's history of accidents involving utilities.[11] For example, during the mid-1980s a PEMEX gas pipeline explosion in San Juanico, a working-class suburb of Mexico City, caused almost one hundred deaths. Moreover, PEMEX had had many accidents in the state of Veracruz. As a result, according to *Política*, "we do not believe that any authority should make a decision of that magnitude without having the assurance that the plant would not fail. The people should have the final word."[12]

The authorities, however, did everything possible to reassure the country that the plant was safe and that "progress" and the national interest should take precedence. For SEMIP, to oppose nuclear technology was "to condemn the country to scientific and technological backwardness."[13] Employing nuclear technology, on the other hand, meant that "the national interests are being defended."[14] SEMIP officials denied that the decision to authorize the operation of the plant had been arbitrary or that the majority of the population opposed Laguna Verde. They instead pointed to its own and the government's efforts to educate the lay population about nuclear energy: for the previous three years, and especially during the last twelve months, technicians from the plant had provided thorough explanations of the project, its security systems, and the advanced technical training they had received. "But unfortunately, the opposition groups rejected the technical arguments and formulated speculations about the dangers that Laguna Verde could cause."[15]

On the same day, October 15, the undersecretary of SEMIP, José Luis Alcudia García, declared:

> I defend what I believe. We are responsible for this project, for this group of actions that the government of the republic and Mexican technicians have taken with seriousness and responsibility. Let us not separate ourselves from progress; let us have faith in ourselves; the federal government under no circumstances would operate something that would fail. Mexican technicians represent a guarantee that Laguna Verde will operate under maximum security in terms of international norms. . . . Let me reiterate that at Laguna Verde there is nothing to hide.[16]

In their attempts to reassure the population as best they could that the plant was safe, government officials revealed that, for them, the debate was not merely about a nuclear power plant but about the trustworthiness of the government and of the nuclear technicians involved with the plant.

In the meantime, representatives of the antinuclear movement continued to counter with the position that the government's decision to run the plant was undemocratic. On October 16 the activists descended on the

Plaza Lerdo in Xalapa, carrying banners and posters denouncing the decision. Protests continued for several days. Antinuclear activists also published obituaries in the local newspapers, expressing their feelings of mourning in the face of the operation of the plant. In these obituaries, they sadly
asked themselves, "The nonviolent, arduous, and constant struggle that
we have undertaken for the last three years, using up all of the resources
at our disposal—have they not been sufficient for the cancellation of this
project?"[17] Another obituary, written by the Madres Veracruzanas, read,
"Yesterday at 6 P.M. . . . the will of the people of Veracruz died upon the
president's decision to operate Laguna Verde."[18] The groups also published announcements facetiously thanking Miguel de la Madrid for "the
historic crime that has been imposed on the people of Veracruz by loading
the reactor at Laguna Verde."[19]

The antinuclear groups also mobilized allies at this time. Cuauhtémoc
Cárdenas quickly traveled to Veracruz to add his voice to the protest. The
bishop of Tuxpan, Veracruz, Ignacio Leonor Arroyo, declared that he and
other bishops from the region had repeatedly asked the government to
respect the will of the people but to no avail.[20]

At various protest sites, Homero Aridjis characterized the decision as
having been made by a "bureaucratic tyranny."[21] He criticized the fact
that the legislature was superseded by the executive branch of government and argued that if the legislators had any dignity, "they would de-
liver a protest" because a visit by a legislative commission had been scheduled for a week *after* the decision to operate the plant was made.

The groups renewed their demand for a popular referendum to assess
the region's opinion of Laguna Verde. Marco Antonio Martínez Negrete
of the Movimiento Antinuclear asked members of the legislature to carry
out a referendum and to amend Article 39 of the Constitution, which
stipulates that sovereignty resides in the people. If sovereignty did indeed
reside in the people, declared Martínez Negrete, then a referendum was
appropriate because that was what the public demanded, and Article 39
should thus reflect this call for referenda when they became necessary.
Several political parties—the PAN, PARM, and PMS—criticized the
government's actions and joined the movement in calling for a referendum on the issue. Only the PPS sided with the PRI by supporting Laguna
Verde and opposing the idea of a referendum on the plant.[22]

As movement leaders continued to make announcements to the press, various groups continued to organize protests in Xalapa and Palma Sola, and the public made its views known. The Madres Veracruzanas of the village of Emilio Carranza declared, "We wish in no way to be disrespectful to the members of our government, but we do wish to be heard, based on fundamental human rights and on rights that nature has conferred to us as mothers. The members of our group reject Laguna Verde. We are not professional environmentalists; we are not pursuing political ends; nor are we retrograde or ignorant people. We are a town of people who love life and our state."[23]

Antinuclear groups also attempted to woo more working-class adherents to join the movement by using the anonymous popular figure Superbarrio to carry the antinuclear message. Superbarrio, who dresses in a mask and tights not unlike a professional wrestler, helps to publicize the problems of the urban poor and working class in Mexico. Genaro Guevara of the Comité Antinuclear of Xalapa revealed that his group was trying to persuade labor to engage in a general strike in order to force the authorities to shut down the plant.[24]

As the antinuclear groups protested, the governor of Veracruz, Fernando Gutiérrez Barrios, traveled to Mexico City to meet with federal officials from the CFE, SEDUE, and the Comisión Nacional de Seguridad Nuclear y Salvaguardas. The governor asked for guarantees from the authorities that there would be no contamination when the plant began to operate. He said, "I want to bring the people of Veracruz a true and trustworthy message that indicates that the interests of the residents of the region will not be harmed."[25]

The decision to operate Laguna Verde was clearly controversial: both the national business daily *El Financiero* and the national left-wing newspaper *La Jornada* published articles raising questions about the wisdom of operating the plant in spite of a weak emergency evacuation procedure and a reactor flawed by design.

While *El Financiero* raised these questions, journalist Guillermo Zamora traveled to Veracruz in search of answers. He discovered a disturbing pattern among various bureaucracies and institutions that ostensibly would be responsible for helping the population in case of a nuclear accident. When he first asked residents of the area what they were supposed to do if an accident occurred at Laguna Verde, none of the residents knew.[26] He received similar responses when he visited the naval hospital at the port of

Veracruz. The PERE, the formal evacuation procedure, designated this hospital as the medical center responsible for tending to victims of a nuclear accident, yet the head of the hospital, Captain Francisco Félix Inzunza, said, "I do not know what to do if the moment arrives. They have not given us concrete instructions."[27] The captain also noted that the hospital had neither the equipment nor the personnel to handle such an emergency. Horacio Fourzán Márquez, head of the Third Naval Zone, where the plant is situated, also admitted that the navy lacked the necessary helicopters, ground transportation, and equipment to handle an evacuation. "We shall use even passenger buses if there is a need. . . . We shall ask the help of fishermen; we have confidence in the solidarity of Mexicans should the sad moment arrive."[28] In his article, Zamora noted that in the case of Chernobyl, the radioactive cloud traveled two thousand kilometers—all the way to Sweden. He surmised that the same could happen at Laguna Verde, but Mexican emergency personnel and institutions were in no way prepared to handle this situation.

The movement's next challenge would be to sway the lower half of the Chamber of Deputies to reverse the executive branch's decision to operate the plant. (The senators had already approved the operation of the plant on October 13.) The president had approved its operation without waiting for the Chamber's lower half to finish its debate on the plant. Antinuclear activists from the state of Veracruz decided to travel to Mexico City in a caravan, which left from Palma Sola on the morning of October 17 and included two buses and six cars, several of which were owned by opposition parties. In the caravan were important opposition party politicians: Cuauhtémoc Cárdenas, Heberto Castillo, Manuel Clouthier, and Rosario Ibarra. As cattlemen of Palma Sola boarded their vehicles, they declared that they were considering withholding their meat and milk products from the Mexico City market in order to publicize their cause.[29]

Upon arrival in Mexico City, the antinuclear activists demanded that a referendum be held to assess the public's opinion of Laguna Verde. This time, however, they insisted that the population should be polled by representatives of the various political parties, not simply by the government.[30] They also immediately organized a protest and declared that they were prepared to engage in more highway blockades.

The debate in the Chamber of Deputies on October 18, 1988, was completely chaotic. The PRI had filled the gallery with *acarreados,* PRI supporters of working-class and peasant backgrounds who had been given gifts of food in exchange for their vocal support during the debate—a common PRI tactic for decades. As the antinuclear groups began to arrive, tensions rose. Interestingly, only certain activists were allowed to enter the building. The Madres Veracruzanas and other middle-class movement members were given free access, but working-class and peasant partici-pants were often denied admittance and had to shout their opposition to Laguna Verde from outside the building. Tensions ran so high that the *granaderos,* special riot police, patrolled the area outside.

Despite the security measures, however, violence erupted. First, anti-nuclear activists tangled with sidewalk vendors; the two groups threw rocks at each other. Then, inside the building, when the PRI acarreados began to shout during the proceedings and to throw food down from the gallery, they were expelled.[31] They were extremely frustrated; they did not understand the debate and wondered why PRI officials were dissatisfied with their performance in the gallery.[32] As they left the building, they im-mediately began to taunt the antinuclear activists who had remained out-side. Although the leaders of the antinuclear groups exhorted their fol-lowers to ignore the provocations, the two groups attacked each other, and several people were injured. PRI supporters also attacked the anti-nuclear groups' vehicles, which were easily identified by bumper stickers expressing support for Cárdenas and opposition to Laguna Verde.[33] Sub-sequently, the antinuclear activists who were outside attempted to enter the Chamber by force, but in the process they broke a plateglass window at the entrance.

In the meantime, the debate continued inside. Antinuclear deputies at-tempted to convince PRI deputies and the legistlative body as a whole that the plant should not be allowed to function and that better security mea-sures should be employed at the plant. A deputy from the PFCRN, Ismael Galán, spoke about the possibility that after an accident, a radioactive cloud, like the one that formed after the Chernobyl accident, could float over the heart of Mexico, doing unimaginable harm to the population and the environment.[34] He proposed that the Chamber of Deputies form a committee to study the effects of nuclear accidents in order to help Mexico decide on proper security measures: "We who are a parliamentary group could exchange experiences with the members of the Green Party, regard-ing its position on the nuclear problem in Germany and with the scientists, intellectuals, and politicians of Sweden, who have participated in the proposal of a referendum to stop the nuclear plants in Sweden."[35] He went on to say that he did not believe that science is neutral, given the roots of nuclear energy in the construction of nuclear weapons.[36]

After almost eight hours of presentations by individual deputies, the debate drew to a close. Javier López Moreno, a PRI delegate, was very blunt: "The decision to put the reactors into operation has been made."[37] He nevertheless also declared that the Chamber was interested in guaranteeing the safety of the population near the plant.

Deputies from all seven parties represented in the Chamber participated in the debate. Only the PPS supported the PRI's position. Modesto Cárdenas García of the PPS maintained that the antinuclear position could be successfully countered with "scientific" arguments.[38] Although he admitted that operating a nuclear power plant involves risks, he was confident that these risks could be minimized through the use of rational thought based on science and technology.

While the Chamber of Deputies was experiencing political tensions and protest, the capital of Veracruz, Xalapa, dealt with its own. For several days after the government's announcement to operate the plant, there were round-the-clock protests in the streets of downtown Xalapa. The Plaza Lerdo was the site of protests for countless groups, from environmental organizations to labor unions. Local, state, and federal government employees engaged in several work stoppages to protest their low salaries and the decision to operate the plant. During these work stoppages, the employees would take to the streets to give support to the environmental groups.[39]

In addition to marches, the antinuclear groups also once again resorted to blockading roads. On October 18, Xalapa fell into a state of chaos because of a two-hour blockade of the major downtown arteries.[40] The newspaper *Política* noted that "whoever wishes to be heard launches himself into the street and blocks the way for vehicles. The dissident groups have learned that this is the way . . . to be heard, and perhaps the problem will be solved. It would seem that among all the types of pressure and petition, only this one has turned out to be effective."[41]

The government's response to the antinuclear campaign varied at this time. On the one hand, various government and CFE functionaries admitted that they had not done a good job of providing the public with accurate information about the plant. As he received a contingent of children who opposed the plant, Alfredo Algarín Vega, the subsecretario de gobierno, declared that there was "a vacuum of information for the public" regarding Laguna Verde.[42] CFE officials also admitted that the evacuation procedure contained certain flaws and would be reevaluated. The director of the CFE, Joaquín Carreón, announced that "we have had errors in form; errors in communication with the population; but fundamentally [the plan] is good. Give us the opportunity, believe in the security of Laguna Verde."[43] He went on to lament certain absurdities in the evacuation plan, such as the recommendation that citizens could protect themselves from radiation by folding a handkerchief sixteen times and holding it to their noses while breathing. "We would not be so irresponsible as to think of fooling you; all of us who work here are Mexicans the same as you, many of us from Veracruz, and I too am from here."[44] CFE officials reassured the population that technicians from the governor's office and from the National Defense Office were analyzing and discussing the emergency evacuation plan and that it would be ready in early November 1988.[45]

Although some government departments and individuals within the CFE attempted to respond to public outcries by reassessing the evacuation plan, the CFE officially continued to publicize its intentions to build more nuclear reactors. On October 18, 1988, Rafael Fernández de la Garza, director of the plant, announced that Laguna Verde was only the first of a series of nuclear plants to be built in Mexico. Four more plants, similar in design to Laguna Verde, would be constructed during the next ten years and were tentatively scheduled to be built in areas such as Soto la Marina and Tuxpan on the Gulf Coast, and in the northern state of Sonora.[46] Thus, the CFE did not appear to be concerned about the antinuclear protestors' arguments or reactions; it intended to keep the Mexican nuclear energy program moving forward.

During the fall of 1988, Mexico secured a bridge loan to help ease the country's economy out of the fiscal crisis that had prevailed throughout much of the decade. The government announced that the bridge loan totaled $3.5 billion and would help to make up for losses resulting from the fall in crude-oil prices. Antinuclear activists noted in cynical tones that the bridge loan was for the same amount as the capital spent on the Laguna Verde plant—$3.5 billion—so they took this opportunity to make yet another argument against nuclear technology: it was too expensive, and Mexico could find better uses for its capital.[47]

As the debate heated up, the people of Palma Sola continued to fear both the Laguna Verde plant and the military detachments still located in the area. The local cattlemen described the villagers' state of mind as similar to that of people living in a war zone. Leaders of the antinuclear movement continued to demand that the government withdraw the military forces because they were increasing the residents' fears.[48] Jesús Darío Rodal Morales, one of the cattlemen, denounced the excessive security measures being taken, which included a patrol of fields surrounding the town. He feared that the soldiers' presence was especially harmful to the children, who avoided the military checkpoints at all costs.

The antinuclear protests in the state of Veracruz continued almost around the clock during the days and weeks after the government's announcement that Laguna Verde would go on-line. Slowly, the arguments began to change from purely technical critiques of the plant to polemical statements against authoritarianism. At a protest in Xalapa many of the leaders voiced opposition to the authoritarianism of the government in operating the plant against the wishes of the local population. Speakers also questioned the legitimacy of the authorities, especially the Chamber of Deputies, and expressed their repudiation of the military in Palma Sola. Juan Marín declared that "if the protest marches aren't enough, we are going to do things that are enough because we do not want that nuclear plant in Palma Sola."[49] Porfirio Muñoz Ledo added, "It has been said with reason that the patience of the public is waning, and just as the reactor is being charged at Laguna Verde, so the reactor of the people is being charged, and for this there is nothing that will hold it back."[50] Cristina Millán, another antinuclear activist, addressed the governor: "We ask the governor to take advantage of this last opportunity to listen to the people of Veracruz—that the plant be stopped until there is a responsible emergency plan that will convince all of us, and that they ask us for our opinion if we want the plant to operate."[51] The movement activists clearly saw the Laguna Verde plant and the faulty evacuation procedure as dangerous projects that had been pushed on the population by an authoritarian regime.

• • •

Miguel Rico Diener, writing in the daily *Uno Más Uno*, argued that the decision-making process in the Laguna Verde case was hasty and displayed characteristics of neocolonial authoritarianism:[52] the lower half of the Chamber of Deputies had not even finished debating the topic when the executive branch announced its decision. Both the CFE and the federal government explained that Laguna Verde *had to operate;* otherwise, the country would soon experience an energy crisis, with central Mexico forced to ration energy within three years. Presumably, Mexico's financial crisis of the 1980s was at least partially to blame for this coming energy crisis; the CFE's budget had been cut by 26.5 percent between 1987 and 1988. Though the CFE did not admit this, it appeared that Laguna Verde had consumed much of the budget, and little had been left over to pursue hydroelectric projects, an area of expertise for many Mexican engineers.

Antinuclear groups had attempted to persuade the government to abandon nuclear energy in favor of plants run by oil and gas because Mexico is a large oil producer. But Rico Diener argued that this plan was not viable for the Mexican government. Mexico's oil production was already at its limit. Moreover, much of the oil was already committed to foreign buyers. In fact, agreements with foreign creditors, such as one signed in 1986, assumed that Mexico would keep producing oil for export in order to meet its debt obligations. Thus, for Rico Diener, authoritarianism and neocolonialism worked hand in hand in this situation. Mexico could not divert its oil production to cover for a canceled Laguna Verde plant because of its obligations to foreign creditors. The government had to ignore the population's opposition in order to avoid an energy and financial crisis.

At the ideological level, the movement emphasized the point that the decision to operate the plant was political. Thus, their demands included a referendum on the nuclear issue. Government officials, however, did not acknowledge this point and defended their decision on practical and technological grounds: Mexico had to pursue "progress," and no referendum was needed to proceed. The antinuclear movement clearly had not been successful in pushing the democratization process further in Mexico.

Although the movement was unable to further the cause of democracy, the antinuclear groups did exhibit New Social Movement tendencies during this period. The NSM literature notes that one of the characteristics of the new movements is a tendency for participants to engage in novel political practices. In the case of the antinuclear movement in Mexico, the activists not only recruited Superbarrio, a figure who participated in numerous protests, but also developed unique rituals to express their displeasure with the operation of the plant. To express their feelings of mourning, the activists published obituaries, and in a protest staged on October 17, 1988, the Madres Veracruzanas brought a coffin to the Plaza Lerdo in Xalapa.[53] Directly in front of the coffin they placed a sign that read, "Here lies the sovereignty of the citizens of the state of Veracruz." Hundreds of protesters gathered around the coffin, which was surrounded by lit candles, and intoned prayers. The newspapers described "a strange manifestation of repudiation for Laguna Verde" in which "upper-middle-class people intoned chants."[54] Though the ritual seemed strange to the reporters, it was very meaningful for the Madres Veracruzanas, who years later still recall it as important in symbolizing the authoritarian political process that had resulted in the operation of Laguna Verde.

5 Palma Sola

The Villagers Experience Repression

Most of the antinuclear groups continued their anti–Laguna Verde activities despite periodic threats from the government, but one group of participants—the villagers from Palma Sola—dropped out. The government did not treat all of the antinuclear groups in the same way. The Madres Veracruzanas were never threatened with repression, whereas members of CONCLAVE reported constant harassment. But perhaps the group that suffered most at the hands of the government was the population of Palma Sola. This chapter concerns the villagers' mobilization and their subsequent withdrawal from the movement.

The villagers of Palma Sola, Veracruz, had not always been opposed to nuclear energy. According to Marta Lilia Aguilar, a local lawyer and antinuclear activist, the people of the region were not even aware that a nuclear facility was being built until after the accidents at Three Mile Island and Chernobyl.[1] The few who did realize what was happening were pleased and excited because of the increased number of jobs that the plant would create. Devoted entirely to agricultural activity, the area offered residents few employment options. Before the accident at Chernobyl, the villagers of Palma Sola had not considered that nuclear technology could be dangerous. Even so, a small number of people were concerned, so they met with former governor Hernández Ochoa, who told them that the decision to locate the plant at Laguna Verde had already been made. Nothing could be done to cancel or relocate the plant. So until the mid-1980s the people of Palma Sola were generally pleased that jobs were available during the construction process at Laguna Verde; moreover, many expressed a sense of pride that such a high-tech plant would be located next to their village.

However, everything changed after the accidents at Three Mile Island and Chernobyl. The Chernobyl accident in 1986 had an especially big impact because of extensive coverage in the Mexican media. At that point, members of the Cattlemen's Association in Palma Sola began to mobilize the villagers. The wealthiest and best-educated residents of Palma Sola, the cattlemen used the resources at their disposal to educate their fellow residents about the dangers of nuclear energy. One of the cattlemen, Jesús Rodal Morales, began holding meetings for the residents, often reading articles to them from *Proceso,* a national political magazine that was critical of Laguna Verde.[2]

The cattlemen and residents quickly made contact with other antinuclear groups, especially the Madres Veracruzanas and the Grupo de los Cien in Mexico City. They realized not only that Laguna Verde was unsafe because of the inherent dangers of nuclear technology but that it posed additional dangers because the evacuation plan was deeply flawed. Shortly after the evacuation procedure was devised, government officials descended on the area, giving away calendars and pamphlets explaining what should be done in the event of a nuclear accident. Instead of instilling a sense of confidence in the residents, however, these materials merely heightened their fears. The government officials' attitude in presenting the evacuation procedure to the people of Palma Sola actually further mobilized the cattlemen. An engineer accompanied by military personnel delivered the evacuation plan in an extremely authoritarian and haughty manner, which unnerved the villagers. The very existence of an evacuation plan also made them instantly aware of the threat that nuclear technology posed. Then, once they looked at the plan, they realized the entire procedure was flawed. The map upon which the evacuation plan was based had not been drawn accurately: entire villages had been left out, some villages had been positioned incorrectly, and nearly impassable narrow dirt roads had been cited as major evacuation routes.[3]

In addition, the pamphlet provided guidelines that the villagers regarded as futile, if not ridiculous, in the event of a nuclear accident. First, it advised that in case of an accident, residents should seal any cracks in their homes to prevent drafts from entering. However, according to Antonio Bretón, a member of Palma Sola's Cattlemen's Association, only about 10 percent of the local population lived in solid structures; the vast majority lived in poor-quality housing.[4] It would be impossible to prevent drafts from entering these homes, many of which were huts made of sticks and thus allowed for free ventilation. Second, the pamphlet advised residents to use everyday items, such as men's handkerchiefs, to protect themselves from inhaling radioactive gases. The pamphlet suggested that residents could protect themselves by folding a handkerchief precisely sixteen times and holding it to the nose while breathing. But perhaps the most chilling advice for the villagers was the warning that in case of an accident, parents should not pick up their children at school because government officials would take the children to shelters. This particular aspect of the evacuation procedure so frightened the residents that many mothers simply sat in the town's main street and cried.[5]

Bretón explained that the Mexican government had not handled the entire episode very well. "They should have come with a team of psychologists rather than with a squadron of soldiers."[6] The callousness shown by the officials gave the villagers the impression that the Mexican government did not care about their welfare. Later, in a meeting with an engineer involved with the plant, the villagers were once again disappointed when he could not guarantee their cattle's safety. The engineer stressed that he was only concerned with human life. As Bretón maintains, however, the villagers not only see the cattle as their main source of income but care deeply about them: "We even name our cows." At the end of the meeting Bretón went up to the engineer and told him, "No tienes argumento [You have no case]."[7]

The Palma Sola villagers' protest strategy and goals were similar to those of other groups in the movement. Like the Madres Veracruzanas, they considered themselves apolitical. They would say that they have never participated in politics and that their only concerns have been to rid Mexico of nuclear energy and to protect their families and ranches. The cattlemen from Palma Sola expressed amazement at finding themselves speaking to high-ranking politicians, something they never imagined themselves doing.[8]

In addition to speaking to government officials, the villagers also participated in demonstrations that included other antinuclear groups. Although the cattlemen always insisted upon their autonomy, they nevertheless wanted to be part of the larger movement. During the late 1980s, at the height of the movement, they worked especially closely with the Madres Veracruzanas. The most dramatic protests in which the residents participated were the three separate blockades of traffic along Mexico's main Gulf Coast highway, which runs through the center of the town.

The people of Palma Sola became more and more alarmed as it became evident that the government intended to open the plant late in 1988. As tensions rose, the government responded by threatening to repress any protests. There had already been a military presence in Palma Sola, but now all branches of the armed forces were mobilized. Almost overnight, four battleships appeared off the coast of Palma Sola and Laguna Verde, and they were visible from the town. Jeeps, trucks, helicopters, and military personnel descended upon the village. Protests were no longer permitted; groups consisting of more than five people were not allowed to congregate, nor were they allowed to leave the area. Palma Sola and the nearby village of Emilio Carranza were in a state of siege. Immediately, antinuclear activists condemned the military presence in Palma Sola. Cattleman Jesús Rodal Morales denounced the exaggerated security measures taken in Palma Sola and told reporters that the town was suffering psychologically. Children and many adults were afraid to walk near the military posts located in the town.[9] Military personnel argued that they were there to secure the highway, but Rodal Morales disagreed with that interpretation, saying, "Highway and road security falls under the jurisdiction of the Policía Federal de Caminos [Federal Highway Police] and not the military."[10]

The turning point for the villagers came on October 21, 1988, when the governor visited the plant (approximately a week after the decision had been made to put it into operation). On that day a group of mothers and children from the two villages marched to the outskirts of Palma Sola to protest the operation of the plant. As they reached the town boundary, however, they were stopped by the military. The mothers explained what they wanted to do, but the soldiers told them they could not leave the village and that they were under orders to prevent anyone from leaving. The soldiers waved their weapons, ready to use force if necessary. Moreover, the military announced that anyone attempting to blockade roads in the area would receive an automatic three-month jail sentence. One of the women told the soldiers, "I know my civil rights. I can leave if I choose to." The soldiers replied, "All we know is that we have our orders. This is a restricted area."[11]

The women decided not to proceed with the march. In the next few days the town was afflicted with what the residents describe as a collective depression. Because the blockade of traffic along the coastal highway was no longer viable, and the military prevented groups from congregating, the population could not decide what to do next. Further, many parents became so fearful of the authorities and of the plant that they stopped sending their children to school. They wanted nothing to do with the government. The period of mobilization came to a close for the villagers in Palma Sola. After this incident, the people of Palma Sola changed. The threat of repression—particularly against women and children—had been too great and too frightening. The villagers could no longer be counted on to participate in the activities of the antinuclear movement. The members of the Cattlemen's Association of Palma Sola, however, remained in the movement for a short while longer.

In the 1990s, antinuclear mobilization has been minimal in the area. A combination of coercion and co-optation used by the government and the CFE have quieted the people living in Palma Sola. The villagers' experiences have led them to believe that government officials do not care about their welfare and are more concerned about money—keeping the plant in operation in order not to lose the large financial investment in it. Further, certain sectors of the village population may have benefited financially from the presence of the plant. Some small-business owners, for example, are experiencing an economic boom because of the increased number of workers in the area, so they support the government's decision to keep the plant running.[12] Moreover, the CFE has provided a new school, clinic, and improved roads, which has pleased some villagers. And finally, many villagers have found employment at the plant. Shutting down Laguna Verde would mean they would lose their jobs; hence, these workers too are pleased with the government's decision.

On the other hand, many of the most ardent antinuclear activists are demoralized and disillusioned. Marta Lilia Aguilar is profoundly disappointed that the villagers have given up. She believes that the people of Palma Sola should not be passive in the face of a dangerous and unjust situation, but she also feels that the local activists' collective spirit has been broken.[13] Antonio Bretón, of Palma Sola's Cattlemen's Association, has grown ill and no longer participates. Several women who were members of the Madres Veracruzanas in Palma Sola caved in to their brothers' pressures to abandon the struggle.

Though many of the residents at Palma Sola withdrew from public protests after their encounter with the military, the cattlemen continued to press the government for concessions. They still demanded that the plant be shut down but also that the government improve the infrastructure of the region and modify the evacuation plan. The regional cattlemen's group, the Unión Ganadera Regional de la Zona Centro, headed by Efrén López Meza, met with Governor Gutiérrez Barrios and offered to suspend its protest activities in exchange for certain concessions from the government: specifically, the construction of new roads, more electrification in the numerous villages in the area, and better information from the government on the status of the Laguna Verde plant.[14] The roads were especially needed because evacuation of the region in the event of an accident would be very difficult, given that many of the roads near the plant are unpaved and narrow. The governor promised only that he would lead a committee that would visit the plant and provide clear information about it. The cattlemen, in turn, complained that the government was spending large sums of money in a public relations campaign for the plant and that this money could be better spent on rural electrification in Veracruz, something that many *campesinos* sorely needed.[15] As of the mid 1990s, however, most of the villagers and cattlemen remain dissatisfied with the government's actions regarding the plant.[16]

• • •

The people of Palma Sola did not have the same political opportunities available to other antinuclear organizations. First, their geographic location put them at a disadvantage. The various government agencies made the plant's security a priority. As a result, the entire Laguna Verde region was under constant surveillance by the Mexican army and navy. This show of force, especially in the confrontation between the female marchers and the military, thoroughly intimidated the villagers. They felt that they could not continue their participation in the face of deadly force.

In addition, social class was a factor. The villagers did not have the class advantages of organizations such as the Grupo de los Cien or the Madres Veracruzanas. In fact, antinuclear activists believed that gender and class factors protected the Madres Veracruzanas from experiencing any repression. The people of Palma Sola did not enjoy these advantages, so they were forced to withdraw from the movement.

All in all, a New Social Movement perspective would be a poor one for viewing the case of Palma Sola. First, the cattlemen's goal cannot be considered to be postmaterial. That is, the cattlemen joined the movement primarily because they were concerned that their agricultural products would be spurned by consumers fearful of radioactivity. Second, the NSM paradigm also does not include an analysis of repression. Yet it was the military's heavy-handed treatment of the villagers that in fact forced them to abandon the movement. Finally, the case of Palma Sola shows that the Mexican government's authoritarian tendencies were still firmly in place.

6 The Antinuclear Movement Exposes Laguna Verde's Problems

On October 15, 1988, the government announced its decision to operate Laguna Verde. In the weeks after the decision was made, the antinuclear groups considered various means to protest. Many groups had not given up on the idea of the blockade of traffic despite the government's threats to repress blockades and arrest participants. They also discussed boycotting payments of electric bills, taking over the CFE offices, and executing collective blackouts on designated evenings in Xalapa and Cordoba in order to make dramatic statements about their opposition to the plant.[1] Antinuclear activists used every opportunity to express their resistance to the nuclear project. For example, on November 2, 1988, the Day of the Dead, a very important religious holiday in Mexico, they met in Mexico City at the Independence Monument to set up an offering to the dead. One activist dressed as Death, in black clothes marked with the outline of a skeleton. The protesters set up an altar with flowers and decorated it with anti–Laguna Verde slogans.

This chapter focuses on the movement's activities and strategies in the period after Laguna Verde went on-line. Overall, the antinuclear groups were extremely active. Not only did they engage in demonstrations and protests but they also took on a new watchdog role: the Madres Veracruzanas, the cattlemen, and a fishermen's cooperative combined forces to monitor the plant's operations. The groups continued to ask for a meeting with the president because they were convinced that only the president could reverse the government's decision. This chapter, then, describes the activists' attempts to publicize the many problems plaguing the Laguna Verde plant in the hope that revealing its technical problems would convince important government officials—including the president—that the plant should be shut down.

The Emergency Drill Scandal

In the midst of the debate about protest tactics, the people of Palma Sola, still anxious about Laguna Verde, now said they suspected that something had gone awry at the plant. According to information given to the Madres Veracruzanas by the villagers of Palma Sola and Emilio Carranza, on Wednesday, November 16, 1988, at approximately 10 P.M., diners at the Fogata Restaurant in Palma Sola heard what sounded like an explosion at the Laguna Verde plant.[2] The next day, at approximately 11 A.M., the villagers saw a great number of soldiers spread themselves throughout a thirty-kilometer area. Two hours later, several residents saw an ambulance arrive at the plant. It carried away two workers—who appeared to be injured—to the Centro de Salud, a local clinic. Shortly thereafter, two doctors and other medical personnel arrived at the plant. A few minutes after that, a helicopter landed and took away two more workers, also presumably injured. As the residents observed these events, they panicked and assumed that a nuclear accident had occurred. Several mothers went immediately to the local schools and removed their children.[3] To allay fears, the CFE announced that the local residents had witnessed a dress rehearsal—a "simulacro de accidente"—not an accident.[4]

When they heard the news about the drill, the antinuclear groups expressed outrage and dismay about how it had been executed without warning local residents. An editorial in the newspaper *Política* argued that the drill had shown once again that the authorities did not care about the local population and did not have the decency to show Palma Sola a little consideration, even though they knew the villagers had been worried for such a long time.[5] The new governor, Dante Delgado Rannauro, immediately announced that all future safety drills would be announced to local residents before proceeding.[6]

Local residents and antinuclear activists speculated that a real accident had occurred, and the fact that the CFE refused to provide details about the drill merely fueled the speculation.[7] José Arias Chávez of the Pacto de Grupos Ecologistas explained that he believed that an explosion due to excessive pressure had occurred in the pipes of the primary cooling system.[8] He speculated that the CFE had been running preliminary tests too quickly in order to get the plant ready for commercial operation, but the plant's cooling system had not responded well.

Initially, the CFE responded indirectly to Arias Chávez's accusation. But later the head of public relations at the plant, Ramón Cosío, countered that the antinuclear groups had begun a new offensive against Laguna Verde through "terrorismo informativo" (information terrorism). Carlos Vélez Ocón, general director of the Instituto Nacional de Investigaciones Nucleares, argued that drills *should* be executed without warning so that the residents would become accustomed to acting quickly in an emergency.[9] The next day, the CFE responded specifically to Arias Chávez's speculations and denied that any explosion had occurred in the cooling system. Spokesmen declared that the Laguna Verde plant was on schedule in the various tests that had to be completed before the plant could operate commercially.[10]

The magazine *Proceso* published an article that gave details about the recent events at Laguna Verde. After investigating the incident, the reporter Guillermo Zamora concluded that the incident at Laguna Verde had not been a nuclear accident but rather a "conventional accident."[11] He also stated, however, that the CFE had conducted an emergency drill on November 18, 1988—two days after a real accident had occurred—in order to cover up the incident. According to unnamed sources working at the plant, the real accident had occurred in an area of the plant responsible for carrying away hydrogen. The problem, according to the workers, was that the pipes carrying the hydrogen should measure 1.5 inches in diameter, but the pipes in use measured only .75 inches. Thus, when the system was tested, the pipes shook violently ("provocó un chicotazo") and hit nearby surfaces, thus causing the noise that the villagers had heard.

Because the incident caused so much discussion, the CFE announced that the Chamber of Deputies would create a commission to investigate. The multiparty group was composed of five legislators—two from the PRI, two from the PAN, and one from the Frente Democrático Nacional (FDN).[12] Problems for the commission surfaced almost immediately. The opposition party members (that is, non-PRI legislators) had wanted to take independent technicians with them to help them interpret the findings at Laguna Verde. The president of the Comisión de Ecología of the Chamber of Deputies, Javier López Moreno, denied this request, however. He maintained that there was no room in the plane for the technicians. The commission arrived at Laguna Verde on November 26, 1988, and was immediately chastised by the head of the plant, Rafael Fernández de la Garza, for listening to "rumors and opinions without basis in fact."[13] To make matters worse, he initially denied the commission access to log notes that they believed would have allowed them to prove whether an accident had or had not occurred.[14]

Eventually Fernández de la Garza relented and allowed the group to see the log notes, but refused to give them copies. Mario Rojas Alba, a physician and member of the commission, said that the notes were disorganized and often illegible. Although he had no training in engineering, he stated he was able to discern from the log that valve 16 was giving faulty information. He and the other members of the commission also inspected the cooling system's pipes, where they noticed that one of the valves showed a fresh coat of black paint. Rojas Alba reported that the director of the plant, Fernández de la Garza, minimized the commission's concerns, categorically denied that any incident or accident had occurred at Laguna Verde, and told the members of the commission that they had fallen into a trap set by the anti–Laguna Verde forces.[15]

In his report to the Chamber of Deputies, Mario Rojas Alba argued that "even if there had been no explosion, there were still indications that an accident occurred, which is being covered up so that Laguna Verde's faulty design will not be evident."[16]

A few days after the commission's visit to the plant, José Luis Alcudia of SEMIP, which oversees Laguna Verde, admitted to the press that "what happened at Laguna Verde was not a nuclear accident but an incident provoked by a problem in the plant's valves. There is nothing to hide at Laguna Verde."[17] Alcudia's announcement put the plant's director, Fernández de la Garza, in a bad light. The PAN and FDN representatives in the Chamber of Deputies called for the resignations of Fernández de la Garza and Juan Eibenschutz, undersecretary of planning for the CFE, for lying and hiding information from the Chamber of Deputies and for risking the lives of millions of Mexicans.[18]

The director of Laguna Verde was thus forced to admit that something *had* happened at the plant, although he dismissed its importance: "An accident did happen, but incidents involving the valves are common at Laguna Verde, where we have four thousand or five thousand valves."[19]

Rojas Alba continued his investigation and discovered that yet another problem existed at the plant—one that had also been covered up. Once again according to unnamed technicians at the plant, two days after the initial accident, on November 18, 1988, the plant's water-cleaning unit failed, causing a stream of water to emerge out of the reactor's vessel. This problem was repaired by plant personnel, who were soaked in the process. The reactor was not operating at the time, and the workers did not appear to have been harmed, but the sources warned that the reactor would have been put into a state of emergency if this problem had occurred during operation.[20]

This incident led opposition deputies to propose that independent technicians be assigned to examine Laguna Verde. In the state of Veracruz, residents sent hundreds of telegrams to President Salinas, urging him to listen to the population of Veracruz and to the antinuclear groups.[21]

Workers Exposed to Radiation

A few months after the first incident at Laguna Verde and while the emergency drill scandal still weighed heavily on movement participants' minds, a new problem emerged. On March 7, 1989, the cattlemen of central Veracruz announced that radioactive gases at the Laguna Verde plant had leaked into the atmosphere.[22] The cattlemen had copies of the plant's official log, which was provided to them by anonymous sources working at Laguna Verde. Efrén López Meza, president of the Unión Ganadera Regional de la Zona Centro (Cattlemen's Association of the Central Region of Veracruz), announced that radioactive emissions had occurred on three separate occasions—on March 1, 2, and 3, 1989. Four workers were exposed to high doses of radioactivity, according to the inside information. In addition, because of high winds on the days in question, up to sixty municipios may have been affected.[23]

The CFE, meanwhile, denied the cattlemen's allegations. A CFE spokesman maintained that routine tests at the plant revealed no problems, and no personnel had been exposed to radiation.[24] He declared that the cattlemen's announcements were unfounded and were intended to provoke fear and alarm in the population.[25]

The Madres Veracruzanas were working closely with the cattlemen during this period, and they responded angrily to the CFE's statement. They demanded that President Salinas do something about the nuclear problem: "There is the proof. Or does President Carlos Salinas de Gortari expect to have cadavers brought to him in order for him to give the order to suspend operations at the Laguna Verde nuclear power plant?" asked María del Carmen Calvo Barquín.[26] Yolanda Lara Sánchez of the Madres

Veracruzanas of Palma Sola denounced this new cover-up by the CFE and announced that the local population would newly mobilize in order to protest.[27] María del Carmen Calvo Barquín suggested that either SEDUE or the Secretaría de Gobernación should test the air quality in Mexico City, as well as in the region surrounding the plant. The Madres Veracruzanas said that they believed the antinuclear movement had enough evidence of the CFE's mismanagement of the plant to warrant an order by Salinas to suspend operations.[28]

As a result of these new accusations, the Chamber of Deputies decided once again to investigate the purported problems at the plant.[29] The Chamber's Comisión de Ecología, led by PRI member Javier López Moreno, also decided to interview the heads of the Comisión Nacional de Seguridad Nuclear y Salvaguardas, the commission that oversees the plant.[30]

The various incidents at Laguna Verde renewed the antinuclear groups' determination to meet with President Salinas. The former president, Miguel de la Madrid, had refused all requests for meetings, but the activists remained hopeful that Carlos Salinas would be more amenable to a discussion of the problems. The Grupo de los Cien organized a march on the presidential palace, Los Pinos, as a public demonstration of their intention. Efrén López Meza of the cattlemen's group insisted that it would ultimately be President Salinas who would decide the plant's fate, so it was necessary to meet with him to discuss the situation.

The cattlemen explained that according to their information, one of the workers, Andrés Vargas, had been exposed to radiation while measuring humidity levels in the statistical panel *(panel de muestreo)*; he had not been warned that radioactive leaks had been occurring, so he had entered the area without the proper protective equipment. López Meza said, "This reflects internal deficiencies because the plant's directors have an obligation to protect the workers."[31]

In addition to engaging in dialogues in the press with the CFE and other government institutions, the antinuclear groups continued their demonstrations and symbolic resistance to nuclear energy. On March 24, 1989, a group of activists from Villa Rica, Veracruz, in the municipio of Actopan, walked to Laguna Verde. When the group reached the plant, they blocked traffic for three hours while they erected a cross in front of it. The participants performed this ritual, they said, in order to pressure the federal government to shut down the plant, especially now that it was experiencing problems.[32] Catholic religious symbolism continued to play an important part in the movement's activities.

Two weeks after repeatedly denying that any new incident had occurred, the CFE admitted that workers had indeed been exposed to radiation. Guillermo Guerrero Villalobos, director of the CFE at the time, admitted that "they were exposed to emissions higher than those established by internal regulations."[33] He revealed that a total of six workers had been exposed to radiation, but gave only two names—Andrés Vargas and Hilario Lagunes. According to Guerrero Villalobos, the accident occurred due to a failure in the alarm system, which did not go off with the detection of a radiation leak.[34] Once again, the cattlemen and the antinuclear movement had pushed the CFE into admitting that problems were occurring at the plant. Guerrero Villalobos minimized these incidents, however. The injured workers, according to his information, had not been exposed to unacceptable levels of radiation.[35]

José Arias Chávez of the Pacto de Grupos Ecologistas and Alfonso Ciprés Villareal of MEM suggested a reason why Laguna Verde seemed to have so many technical problems. According to the two activists, corruption was largely responsible for the plant's failures.[36] Low-quality pipes, valves, and spare parts had been purchased, thus causing the frequent problems and delays in the plant's operation. Arias Chávez and Ciprés Villarreal counted five accidents since the government had authorized the plant's operation, despite the fact that the reactor was functioning at only 5 percent of capacity.[37] The two concluded that an independent inspection of the plant was needed in order to ascertain whether it was safe.[38]

Stoppages and Radioactive Leaks at Laguna Verde

Throughout most of 1989 the Laguna Verde plant experienced constant stoppages as technicians eased it through various tests before allowing it to operate commercially. The pattern continued. At every turn, the antinuclear activists, especially the cattlemen, would break the news to the press of irregularities at the plant, such as stoppages and the dumping of radioactive water. The CFE would deny that problems existed, only to admit later that something had occurred but that it was not as serious as the antinuclear activists had made it appear.

Between April and August of 1989 the plant experienced sixteen stoppages; many of these occurred automatically as certain systems in the plant failed. For example, the ninth stoppage, on April 28, 1989, took place because metallic residues were detected in the water feeding the reactor.[39] A few days later, on May 4, 1989, the reactor experienced another automatic stoppage because of irregularities in the instrumentation in one of the control systems.[40]

Although plant officials generally admitted that these stoppages were occurring, they would not admit to radioactive leaks into the atmosphere or into Gulf waters. However, the cattlemen's group, along with a fishermen's cooperative, were ready to denounce these incidents as they occurred. Efrén López Meza of the cattlemen's organization declared, "We have precise information that on Friday, March 24, a leak of radioactive water occurred and reached the ocean. The contamination was minimal, but it occurred."[41] The cattlemen believed the unnamed technicians' assertions that many of these irregularities were actually caused by human error rather than by purely technical breakdowns.

The cattlemen's denunciations were reiterated by Jesús Aguilar Enríquez, head of the aquarium of the state of Veracruz. According to Aguilar Enríquez, the Laguna Verde plant was dumping hot water into the Gulf of Mexico.[42] Though this water had been treated before being released, its high temperature had caused various fish kills, which in turn affected the food chain so that other species of fish abandoned the area. According to Aguilar Enríquez, the Gulf waters surrounding the plant were now devoid of aquatic life.

The CFE declared that Laguna Verde was responding normally to the tests the engineers were performing at the plant. The director of the plant, Fernández de la Garza, explained that the stoppages at Laguna Verde were normal, given that the reactor's operating strength was being increased very slowly.[43] At each stage the reactor was stopped in order to examine its behavior. He also explained that stoppages were supposed to occur automatically when any of the security systems detected a possible anomaly, but "this does not mean that there is grave danger in the installations."[44] Rather, the security systems were so sensitive that the reactor was stopped even when a minor possible problem presented itself. Fernández de la Garza was very critical of the antinuclear movement. "I agree that society should demand the strictest security measures, but I don't agree with attempts to create alarm, to lie, or to create panic among the residents of Veracruz. I feel that there is pain and bad faith in that."[45] He also assured the environmentalists that the CFE would provide them with any information about the plant in order to calm the public's fears.

The Madres Veracruzanas were not satisfied with the CFE's declaration, however, and told the press that the CFE was still withholding information, especially concerning radioactive leaks. "The radioactive leaks continue, and we have the proof; it is in the changes in the environment."[46] The mothers pointed to the case of the fish kills in the waters surrounding the plant, saying that flora and fauna had similarly been damaged in the Chernobyl area after the 1986 accident.

Later, in May 1989, the Movimiento Ecologista Mexicano, the Grupo de los Cien, and the Pacto de Grupos Ecologistas also attacked the CFE, saying that the high price paid for the plant, with its obsolete technology, demonstrated the ineptitude of the bureaucracy.[47] The three groups added that the various safety drills executed by the CFE had been a failure despite the CFE's declarations about their success: the drills did not conform to the exact evacuation procedure, the PERE.

Ten days after the various attacks by the antinuclear groups, the CFE admitted that radioactive water had indeed been dumped into the Gulf of Mexico.[48] During its weekly report, it announced that on July 8, 1989, the plant had dumped 250 liters of water with slightly higher radioactive levels than permitted by established standards. The CFE still maintained, however, that this incident had not harmed the environment because the total dosage permitted in the environment was not exceeded.[49]

The Madres Veracruzanas disagreed with the CFE's interpretation of the effects of the incident, however. They countered that their anonymous sources revealed that unusually high levels of radioactivity from leaks had been detected as far as five hundred kilometers from the plant.[50] Rosalinda Huerta, speaking for the Madres, warned that continued radioactive emissions from the plant could affect the food chain by contaminating certain agricultural products as well as seafood.

The CFE responded by trying to persuade the public that it did not wantonly dump radioactive water into the Gulf. Officials declared that the water was tested first through chemical analyses using radioisotopes in order to ascertain the water's levels of radioactivity "to guarantee that [the levels] were within the limits set by norms in this [field]."[51] These tests were performed by the plant's own laboratory and were sanctioned internationally, according to CFE officials. They also explained that despite the dumping of contaminated water into the Gulf, ocean waters did not reveal a high level of radioactivity, nor did there appear to be any alarming changes in the environment in the surrounding area.

One Anonymous Source Expelled from Laguna Verde

Shortly after the CFE tried to reassure the public about the plant's safety, it made a move that stunned the antinuclear groups. An engineer at Laguna Verde, Miguel Angel Valdovinos Terán, was abruptly relieved of his duties at the plant and transferred to an obscure position within the CFE in northern Mexico. The CFE accused Valdovinos Terán of providing the antinuclear groups with confidential information about the plant.[52] This information had then been used by the cattlemen, fishermen, and Madres Veracruzanas in their announcements about technical problems at Laguna Verde. Valdovinos Terán had been head of the plant's laboratory and thus had access to information concerning levels of radioactivity.[53] The antinuclear groups did not admit publicly that Valdovinos Terán had been their anonymous source, but Eduardo Gómez Téllez, head of a local fishermen's cooperative, declared that he had sought protection for Valdovinos Terán from the undersecretary of Gobernación, Emilio Rabasa Gamboa. After leaving his post, Valdovinos Terán said that he believed that SEMIP, the government agency overseeing the plant, was attempting to cover up the fact that radioactive leaks were occurring almost daily at the plant. During July 1989, he claimed, the plant had dumped ten million liters of water containing uranium into the Gulf of Mexico. He said that the Unión Ganadera (the cattlemen) had test results from the plant that showed that even when the plant was not operating, it was still leaking radioactive emissions.[54] Eventually, he publicly admitted that he had indeed been the anonymous source of information for the antinuclear groups. Now that he was gone, it would be more difficult for the groups to play their watchdog role.

Marco Antonio Martínez Negrete, a leader of the antinuclear movement, argued that the CFE was engaged in an identifiable pattern of deceit and cover-up. Based on the November and March incidents, he pointed out that in each case (1) independent sources first sounded the alarm about a problem; (2) the CFE immediately denied the accusations, citing bad faith on the part of the environmentalists; and (3) high-level bureaucrats then confirmed the original accusations.[55] Martínez Negrete noted that this pattern was troubling for those concerned with the safety of the Mexican population because it showed the CFE's willingness to maintain secrecy about accidents and also because it indicated that the CFE would probably continue to engage in this behavior in the future.

Fishermen Denounce Laguna Verde

Throughout the second half of 1989 the movement continued to focus much of its attention on Laguna Verde's defects. In late August 1989 the fishermen's organization, Sección de Cooperativas Pesqueras, complained about the effects of the ten million liters of radioactive water that had been dumped into the Gulf of Mexico in June and July 1989.[56] The fishermen took their complaint to the head of Gobernación, Fernando Gutiérrez Barrios, also the former governor of Veracruz: "The results [of what is happening there] are alarming for fishing interests in general and for the cooperative sector in particular. . . . We are in agreement that technical progress should be incorporated into the country, but it should provide the guarantee of not causing problems in the productive sectors of the region."[57] The fishermen's position thus was more moderate than that of the other major antinuclear groups. They were willing to coexist with Laguna Verde—as long as the plant was operated correctly (according to strict international standards) and did not spew contaminated water into the Gulf. The Madres Veracruzanas and Grupo de los Cien were more strident, however. They were deeply suspicious of nuclear technology and believed it was *inherently* dangerous. These groups wanted Laguna Verde to be shut down or converted into a conventional plant: there was no room for compromise.

Upon hearing that Laguna Verde had stopped operating for the sixteenth time, the cattlemen and fishermen jointly announced that they planned to meet with CFE officials in Mexico City, where they would demand that Rafael Fernández de la Garza be removed from his post as director. The subsecretario de gobierno (equivalent to a U.S. lieutenant governor), Rafael Hernández Villalpando, acknowledged that a meeting between the CFE and the cattlemen and fishermen would indeed be held. He reassured the press that the government would close the plant if the cattlemen and fishermen provided convincing evidence. The meeting was held on September 4, 1989. The cattlemen were represented by Dr. Ricardo Bello González, and the fishermen by Eduardo Gómez Téllez. Among the government's representatives were Rafael Hernández Villalpando and Armando Méndez de la Luz, director of Protección Civil of the Secretaría de Gobernación. Also in attendance was Juan Eibenschutz, "father of Laguna Verde" and subdirector of the CFE. The meeting was tense: the cattlemen and fishermen presented the information about the plant's radioactive leaks into the Gulf and the atmosphere, and they gave the officials a study done by researchers at the Colegio de México, which claimed that the evacuation procedure had serious flaws. In a raised voice, Eibenschutz, speaking for the CFE, contradicted the activists. The groups accused the plant of dumping seven million liters of contaminated water into the Gulf per month, but the CFE admitted to dumping only eight hundred thousand liters per month. Eibenschutz added angrily, "We can throw away all the water we want. There are no limits."[58] He accused the cattlemen and fishermen of "victimizing us with their lies in the press; they're destroying us."[59] He demanded that the groups provide concrete proof to back up their claims. The groups' adviser provided data, which included test results showing high levels of radioactivity in the atmosphere, as well as in waters surrounding Laguna Verde. Eduardo Gómez Téllez, head of the fishermen's cooperative, also added that high levels of radioactivity had been measured in an exclusive neighborhood near Laguna Verde, precisely where plant director Rafael Fernández de la Garza lived.[60] The meeting grew tense, with tempers raised on both sides. The undersecretary, Hernández Villalpando, had to intervene several times. Although Eibenschutz acknowledged toward the end of the meeting that the groups' data on water contamination was correct, he accused the two groups of lying to the press.[61]

The cattlemen and fishermen declared that they were disappointed with the meeting because no indication was given that the CFE would change the way it operated Laguna Verde. The groups feared that their agricultural products, such as citrus fruit exported to Japan and the United States, would have no buyers if the radioactive emissions continued.[62]

The Grupo de los Cien subsequently spoke to the press about the fishermen and cattlemen's evidence, stating that they also had inside information that pointed to problems in the way that Laguna Verde was run. Ofelia Medina, Feliciano Béjar, and Homero Aridjis described situations in which plant personnel had gone into contaminated areas of the plant without protective clothing.[63] They had evidence indicating that more whistle-blowers had been fired. Javier Barrera Quirarte, superintendent of technical support, had been fired for writing a document, "Diagnóstico sobre la situación de Laguna Verde" (Diagnosis of the Situation at Laguna Verde). This document, signed by mid- and high-level technicians at the plant, had been written for high-level officials and suggested that the plant's operation be delayed until personnel could be properly trained and further tests could be performed on the plant. It also indicated that because of the many delays in the plant's construction, dating back to the 1970s, the spare parts needed at Laguna Verde were no longer usable because they had not been stored properly on the premises. Thus, when Unit I needed spare parts, they were either purchased, at a cost of $800 million, or taken from Unit II, which was not yet on-line. The cost of spare parts and other materials represented 23 percent of the total cost of the plant.[64]

In the midst of the dialogue between the fishermen, cattlemen, and CFE and government officials, the Secretaría de Marina (the navy) announced—in response to the fishermen's claim that they were no longer allowed to fish in a five-kilometer square area between Palma Sola and Laguna de Tamiahua—that there was no contamination of Gulf waters and denied that the fishermen had been prohibited from fishing in this area.[65] Admiral Mauricio Schelke Sánchez said that he was nevertheless ordering technicians from the Dirección General de Oceanografía to write a detailed report on the matter in order to calm the fears of the people of Veracruz. The cattlemen and fishermen maintained their position and demanded that Rafael Fernández de la Garza be removed from his post as the head of the plant. Francisco Espinoza of the cattlemen's group said it was clear that Fernández de la Garza had been covering up information about the plant's defects; he suggested that Fernández de la Garza had even hidden this information from Juan Eibenschutz: "Juan Eibenschutz did not know that twenty-two million liters of contaminated water had been poured into the ocean; he had not communicated this to the subdirector of planning for the CFE."[66]

Heads of government agencies continued to counter the fishermen's accusations that contaminated water was being dumped into the Gulf. The SEDUE state delegate for Veracruz, Roberto Rodríguez Galera, denied that Laguna Verde had dumped radioactive water into the Gulf. Governor Dante Delgado Rannauro made similar statements, discounting the fishermen's allegations but admitting that plant officials had not done a good job of keeping the population of Veracruz apprised of what was happening at the plant.[67] Finally, Dr. José Rodríguez Domínguez, a member of the planning group for the evacuation plan (PERE) and head of Salud Pública (Public Health Department) of the state of Veracruz, argued that the plant was not leaking contaminated water and that it possessed extremely advanced technology, equal to the best in the world.[68] He asserted that PERE officials would warn the population if any problem arose.

Though government officials flatly denied that Laguna Verde was harming the environment, the CFE nevertheless promised the public that it would change procedures at the plant. The director general of the CFE, Guillermo Guerrero Villalobos, promised to "undertake organizational changes at Laguna Verde"[69] but asked for time to make the changes because they could not be made overnight. He did not indicate the exact nature of these changes, however.

Thus it appeared that although government officials refused to grant the antinuclear groups' demand to close Laguna Verde, the groups were nevertheless effective watchdogs. Indeed, Governor Dante Delgado Rannauro acknowledged in a meeting that the Madres Veracruzanas "had played an important role in ensuring that safety mechanisms were improved and enhanced."[70] The governor publicly supported the Madres' actions and offered to assign a technician to observe the plant in order to ensure that no radioactive leaks were occurring. He also arranged for Rafael Hernández Villalpando, the subsecretario de gobierno for the state, to receive any complaints from the Madres in order to direct them to the proper government authorities.

Sabotage

By late September 1989 the plant had experienced eighteen stoppages since going on-line. High-level officials began to consider the possibility of sabotage.[71] Officials from Laguna Verde and the Comisión Nacional de Seguridad Nuclear y Salvaguardas (CNSNS) decided to restrict access to the reactor and turbogenerator buildings because they believed that many of the problems could have been caused by workers at the plant.

The prime suspects for causing sabotage at the plant—at least in the eyes of the CFE and CNSNS—were the antinuclear groups. If technicians such as Valdovinos Terán were passing information about the plant to antinuclear groups, then it also seemed plausible that workers at the plant might have been persuaded by the groups to engage in sabotage. When they heard this latest accusation, the antinuclear groups immediately cried foul. At a meeting of forty groups in Teziutlán, Puebla, they declared that the CFE and authorities at the plant just could not accept the fact that the plant was deeply flawed, so they were blaming the problems on fictitious sabotage.[72] Mariano López of the Grupo Antinuclear Arcoíris of Xalapa argued that officials at the plant were trying to absolve themselves of responsibility at Laguna Verde.[73] No antinuclear group believed that sabotage was responsible for the plant's many stoppages.

Opposition parties within the state legislature of Veracruz also took the CFE officials to task. The Partido de la Revolución Democrática (PRD) and PAN legislators demanded that Director Fernández de la Garza inform the population about what was happening at Laguna Verde. On September 24, 1989, Ismael Cantú Nájera of the PRD said that he was especially concerned about a rumored new spillage of twenty-two million liters of radioactive water a few days before, as well as about the report of sabotage at the plant.[74]

During the last week of September 1989 leaders of the antinuclear movement demanded that the office of the Procuraduría General de la República (attorney general) investigate the possibility that sabotage had occurred at the plant.[75] The activists did not truly believe that sabotage had occurred, but publicly they argued that because sabotage is a crime, the Procuraduría General should investigate. They were calling the CFE's bluff.

The head of the CFE, Guillermo Guerrero Villalobos, denied that sabotage could exist at Laguna Verde. He declared Laguna Verde had no problems, whether safety related or otherwise. Although he denied that sabotage had occurred, he also stated that "some valves were deliberately opened," but he did not know if this had been due to human error.[76]

Either way, he said, the matter was under investigation. Thus, it appeared that once again the antinuclear groups were playing their watchdog role effectively. When they challenged the notion of sabotage—introduced by the CFE itself—the authorities backed down and admitted that sabotage was not the cause of Laguna Verde's many stoppages and problems.

•　•　•

The antinuclear movement was extremely active in the year and a half after the president's decision to begin to operate Laguna Verde. The groups attended countless meetings with government and CFE officials in an effort to persuade them that the plant had serious structural defects and that it was emitting radioactivity into the atmosphere and the Gulf of Mexico. They thus took on a new watchdog role—publicizing Laguna Verde's many technical problems and procedural violations. Eventually, however, the CFE fought back: Miguel Angel Valdovinos Terán, an engineer working at Laguna Verde, was abruptly relieved of his duties and reassigned to an obscure post in northern Mexico when the CFE accused him of providing confidential information to the antinuclear activists. Certain government officials were thus active in attempting to thwart the movement's efforts to shut down the plant.

In their activities in this year and a half, groups such as the cattlemen and fishermen were clearly motivated by material concerns. The fishermen said that they were concerned about the government's position of indifference on the dumping of contaminated water into the Gulf and complained that this lack of regard for the Gulf meant that their livelihood was in jeopardy. Both groups wanted to protect the sources of their income and thus do not fall neatly within the parameters of the NSM paradigm, which in its focus on issues of identity clearly does not address the fact that a movement's participants can exhibit characteristics of a New Social Movement while also embracing materialist goals.

7 Movement Politics

During the period immediately following the government's decision to operate Laguna Verde, the antinuclear groups believed that they had to strengthen the movement somehow. Some of the activists believed that the movement would be more effective if the various small groups (such as the Pacto de Grupos Ecologistas and the Grupo Antinuclear de Xalapa) were combined into one larger group. Many of the small groups indeed joined together and formed the Coordinadora Nacional Contra Laguna Verde, known as CONCLAVE. The government, however, seemed determined to weaken the movement by offering political positions to movement leaders. Activists were thus constantly suspicious that fellow members would be co-opted. The Madres Veracruzanas were especially worried about this possibility and ended up expelling their unofficial leader, Rebeka Dyer. Despite these problems, however, they continued their antinuclear activities and secured a meeting with President Salinas.

The Founding of CONCLAVE

On February 18, 1989, the antinuclear movement held a national meeting. The convention—the Primer Encuentro Nacional por la Vida y por la Paz (the First National Meeting for Life and Peace)—was held in Xalapa and was attended by antinuclear activists and groups from as far away as Durango, San Luis Potosí, and Jalisco.[1] At this meeting, CONCLAVE was formed, and it would come to play a prominent part in the movement in the next few years. Though CONCLAVE did not have a formal, hierarchical structure, activists Pedro Lizárraga and Juan Marín quickly came to play leadership roles in the group.[2] The participants at the national convention demanded the resignations of Juan Eibenschutz and Rafael Fernández de la Garza, the withdrawal of the military from Palma Sola, the creation of a popular tribunal to judge and sanction the officials and technicians running the plant, and the reform of Article 39 of the Constitution (which maintains that sovereignty resides in the people) to enable the execution of a referendum.[3] Finally, the participants agreed to organize a national march against Laguna Verde, from Mexico City to Palma Sola on April 30, 1989.[4]

The formation of CONCLAVE revitalized the movement. Many of the antinuclear organizations had slowly lost members. The Grupo Arcoíris, for example, in reality consisted of one member—Mariano López. CONCLAVE thus brought together small groups like this and allowed the members to join forces by sharing resources and information. The Madres Veracruzanas, ever vigilant of defending their autonomy, chose not to join CONCLAVE, however. They supported and worked alongside CONCLAVE but refused to merge with it because they feared that their particular point of view would be compromised by joining forces.

The Cattlemen's Leader Defects

Even as the antinuclear groups lobbied the government to shut down Laguna Verde, they also tried to keep their distance for fear that the state would co-opt their leaders. Fernando Jácome, one of the most active members of the movement, had already been elected *presidente municipal* (mayor) of the town of Coatepec as a member of the PRI. Many of the activists were angered by Jácome's decision to join the political system. The Madres Veracruzanas were particularly vocal on this matter because they had always expressed hostility toward political parties and traditional avenues of political participation. They were dismayed again when the PRI announced that the head of the Cattlemen's Association, Efrén López Meza, would run as a PRI candidate in upcoming state elections. Many antinuclear activists saw this co-optation as an attempt to weaken the movement. Marco Antonio Martínez Negrete theorized that the PRI and the central government were worried that the mobilization of the antinuclear groups would translate into votes of opposition in the next elections. Therefore, by convincing López Meza, an important leader of the movement, to run as a PRI candidate, the party would protect itself in the elections.[5] Martínez Negrete was confident, however, that no matter the candidate, the population of Veracruz would see through the ploy and would continue to oppose the government's policies on nuclear energy.[6] Nevertheless, Fernando Jácome and Efrén López Meza's new affiliation with the PRI was very disturbing to participants in the movement. López Meza's action greatly demoralized the Cattlemen's Association—so much so that later the cattlemen would drop out of the movement for good.

The Madres Veracruzanas Expel Their Leader

The cattlemen were not the only group to experience feelings of uneasiness about the possible co-optation of important leaders. The Madres Veracruzanas actually expelled one of their members because they suspected that she had ties with the government. When the member's husband accepted a position with Obras Públicas (Public Works) in Coatepec, Veracruz, the Madres interpreted the job offer as an attempt on the part of the government to compromise Rebeka Dyer's antinuclear actions.[7] Dyer disputes this allegation: "My husband teaches at the Universidad Veracruzana; he is an engineer. Since he was on sabbatical, he was asked to be head of Obras Públicas of Coatepec. He had felt himself to be underutilized at his university position; he wanted applied work as well. So he decided to accept."[8] She says that this decision had nothing to do with her own work in the Madres Veracruzanas, however.

Problems between the Madres Veracruzanas and Rebeka Dyer came to a head when Dyer received an invitation to the governor's Informe de Gobierno, which would take place on November 30, 1988. Patricia Ortega, another member of the Madres Veracruzanas, and two cattlemen also received invitations. The Madres, however, decided that they would send no representatives to the event because they did not want to appear to be too close to the government. Rebeka Dyer disagreed; she thought they should go in order to maintain contact and dialogue with the government. The Madres maintained that the new governor, Dante Delgado Rannauro, had invited them only to try to figure them out, but Dyer thought they should accept anyway in order to present their antinuclear views. The other mothers were thus extremely unhappy with Dyer's subsequent behavior. In retrospect, Dyer says, "I decided to treat the invitation as a personal one, addressed to me. The mothers were against it, but I decided to go. I figured that somebody had to dialogue. Not everyone should be on the outside looking in. I wanted to present the governor with antinuclear views."[9]

The Madres and other antinuclear activists stood outside the Hotel Xalapa as Rebeka Dyer, government officials, and other invited guests went inside to have breakfast with the governor. Dyer was not the only member of the movement to attend: several cattlemen were also present at the breakfast. Dyer wore the red ribbon symbolizing her affiliation with the Madres and feels that, overall, her intentions were good. She explains, "Everyone can play different roles in an organization. Unfortunately, the Madres interpreted my action in a negative way."

Matters between the Madres and Dyer soon grew worse. December 3, 1988, was Rebeka Dyer's birthday, and some of her siblings who live in the United States had come to Xalapa to attend her birthday party. That Saturday, the day of the weekly *plantón* (protest), the Madres had planned a surprise party for her, but Dyer did not attend the plantón because she did not want to leave her siblings, who were visiting for only a few days. She says that perhaps she should have gone anyway. The Madres were offended that she did not attend the plantón/party and saw her action as another indication that she was moving closer to the government. Dyer says, "I failed them emotionally."[10] But Sara González, another member of the Madres Veracruzanas, says that "it was a whole series of things, not just one event"[11] that led the Madres to reject Rebeka Dyer and to ask her to leave the organization.

After her expulsion, Dyer went to see Fernando Gutiérrez Barrios, secretario de gobernación (also the former governor of Veracruz), and was received in fifteen minutes. It had taken the Madres two and a half years to obtain a meeting with Manuel Bartlett, the previous secretary. For Dyer, this difference meant that increased contact with the government had led to more dialogue. The Madres Veracruzanas, however, suspected that it was yet another sign that Dyer had grown too close to the government and had been co-opted.

According to Dyer, many new people joined the groups in the movement in the period after she was expelled, and the new people brought new agendas. CONCLAVE, she argued, had far-ranging political goals beyond mere opposition to Laguna Verde. After the July 6 presidential election, the Foro Cívico de Xalapa was formed, but in Dyer's eyes it was really just a Cardenista front. She shared the Madres' belief that all antinuclear groups should keep their distance from political parties: "The same people [from the Foro Cívico] now are in the Coordinadora [CONCLAVE]. And the Madres worked with the Coordinadora. That's when I decided that enough was enough."[12] She was dissatisfied with the fact that as the movement expanded, opposition to Laguna Verde became only one of many goals for some participants in the movement.

Dyer also complained that after her expulsion the Madres began to say that the government had paid for her new condominium and had given jobs to her son and husband. "The mothers cut me off completely. But I would do it again in the same way. I fought hard with the new CFE director, in front of the governor. Now I am not any less committed to my antinuclear ideology, but I'm in a new phase of participation."[13]

Dyer still wanted to participate in the antinuclear movement, so she created another mothers' group—Mujeres Activas Mexicanas Antinucleares (MAMA). Three of her friends joined because they all felt rejected by certain groups in the movement. Indeed, all of these activists refused to work with CONCLAVE because they believed it was too strident. MAMA never got off the ground, however. A march on December 22, 1988, was a disaster; MAMA had wanted ten thousand women to attend, but only two hundred marched.

As with the members of many New Social Movements, the members of the Mexican antinuclear power movement came to think of their organizations as more than mere vehicles for social change: participants were tied to their groups by friendship and to the movement by constant personal interaction.[14] Thus, for Rebeka Dyer, her separation from the Madres Veracruzanas was a personal loss.

The Madres Meet with the President

Though Rebeka Dyer was no longer with the group, the Madres Veracruzanas continued to work at a fever pitch. In early November 1989 they complained that they had asked countless times for a meeting with President Salinas, yet no invitation had been forthcoming, so they asked Governor Dante Delgado Rannauro to intercede on their behalf.[15] President Salinas was scheduled to come to Veracruz later in November, and the Madres were anxious to take advantage of this opportunity.

On November 8, 1989, the Madres received word that President Salinas had agreed to meet with them. Sara González, a member of the group, announced that the meeting was to take place on November 23 at the Hotel Xalapa.[16] The Madres were ecstatic. They had requested meetings with the president on 250 occasions throughout a three-year period and

and now were finally successful.[17] President Salinas refused to meet with the other antinuclear groups.

Ironically, the announcement that the president would meet with the Madres provoked a crisis within the group.[18] The subsecretario de gobierno, Rafael Hernández Villalpando, who arranged the meeting, told the Madres that they had been allotted a limit of eight persons to attend the meeting with the president. During the two weeks before the meeting with the president, the Madres had a very difficult time choosing the eight representatives. Because the Madres had consciously decided to forego a formal structure and elected positions, the decision had to be made by consensus. Initially, they selected a group of eight simply by shouting out their names during a meeting. Subsequently, however, one faction became increasingly angry because one particular member had not been selected. Individuals in the group discussed this matter over the telephone for the next two days until one of the eight volunteered to step down. In this way, the previously slighted Madre could join the eight selected to meet with the president.

This incident illustrates some of the weaknesses in organizations analyzed by the New Social Movements literature. These mostly grassroots organizations tend to be nonhierarchical and usually have a fluid, informal structure, which allows them to respond in a creative and spontaneous fashion to the problems they are trying to address. As Alberto Melucci has argued, these small groups tend to be united not only by their cause but also by their everyday, affective relations with each other (see chapter 1).

New Social Movements scholars, however, often romanticize this informal, nonhierarchical structure and do not analyze the problems that can emerge from it. Ordinarily, an organization asked to send representatives to a meeting might simply send the elected officers. The Madres, however, have no formal, elected leaders, and they are quite proud of this fact. As this instance demonstrates, the lack of a more formal structure can sometimes cause serious crises within groups: it nearly tore the group apart in the weeks before the meeting with President Salinas.

Despite the problem described above, the Madres' meeting with President Salinas on the night of November 23, 1989, went smoothly. The eight members were confident and poised in their conversation with the president: many of them later commented that they had come a long way since the early days, when they had no political experience. During the half-hour meeting, the Madres presented their views; each representative explained different aspects of the danger posed by the plant. A physician explained the dangers of radioactivity for the population's health, and a psychologist explained the psychological and social harm that could occur after an accident. The president and the governor of Veracruz listened intently, though both interrupted often. At the end of the meeting, Salinas responded by saying that technical consultants had advised him that a conversion of a nuclear plant to a combustible plant was not feasible. However, he promised the Madres that Laguna Verde would be subject to an independent study and that the nuclear plant's fate would be decided by the results of that study. The Madres, though disappointed that he had not agreed to close the plant immediately, were nevertheless pleased that the plant would be investigated by outside technical experts. The president told them that he would announce the members of the inspection team at a later date.[19]

Although overall the Madres were pleased with the results of the meeting, they continued their weekly protests in Xalapa's main square. In early December of 1989 they held a rally to remind the president that they were taking him at his word and expected the independent audit to take place soon. They recognized that they had a stake in the selection of the members of the panel who would inspect the plant, so they offered to pay the fees of an independent technician to oversee the activities of the inspection team.

The president had been vague about a timetable for the study, and no more was heard about it until February 1990. At that time, the director of the plant, Rafael Fernández de la Garza, declared that he had every confidence in the plant and that it would be running at 100 percent capacity by 1993. In response, the governor of Veracruz declared that the president had informed him that the study would take place before the ultimate fate of the plant was decided[20] and that Fernández de la Garza was therefore not authorized to make such statements. The governor also said that the antinuclear activists were within their rights in continuing to protest against the plant as long as they staged their demonstrations within the limits of the law and did not harm individuals or property.[21]

By early March 1990, the CFE began to outline the procedure for the selection of the study team promised by the president. The director of the plant announced then that the independent study, in order to be considered impartial, should be conducted by an independent body with no connections to Laguna Verde.[22] An international search would be launched to select the members of the study team. Details about how the search would be conducted and who would select the team members were not provided.

As the antinuclear activists waited anxiously, a scandal broke out. A fax sent by the director of the CFE to the governor of Veracruz was leaked to the media.[23] It indicated that the selection process would be rigged, essentially to ensure a positive review so that the plant could keep operating. "A formal competition would have been too complicated," it stated. Instead, a U.S. consulting firm would be hired, but it would look as if it had won the competition fairly, "with the object of eliminating possible criticisms."[24]

The daily newspaper *Política* revealed that the fax was sent by mistake to a wrong fax number. The recipients of the fax decided to make the information public and delivered the document to the newspaper's offices. These individuals told *Política* that they were motivated by a desire to protect the interests of the people of Veracruz. They decided to remain anonymous.

According to *Política*, the CFE had rigged the study, but clearly the governor of Veracruz and the heads of such important agencies as SEMIP and SEDUE—officials at very high levels of government—were also involved.[25]

Although outraged by this news, the antinuclear activists still held a faint hope that the study might be conducted impartially. CONCLAVE recommended two engineers who might be included in the investigative team. They suggested Robert Pollack of the United States and Marco Antonio Martínez Negrete of UNAM, both known for their antinuclear views. However, the government announced that a Spaniard, Manuel López Rodríguez, would head the team. None of the people proposed by the antinuclear groups was included.[26]

Although the antinuclear groups sought to meet with government officials to express their displeasure with the selection process, only the Madres Veracruzanas were granted such an audience. In a meeting with Governor Dante Delgado Rannauro, they indicated they had little confidence in the honesty of the study. They proposed that an additional four technical specialists be included in the investigative team, but the suggestion simply "hung in the air."[27]

The governor assured the Madres that the plant was not yet operating commercially, something that the antinuclear activists insisted should not happen without guarantees that it was safe. The governor insisted he would be the first to protest Laguna Verde's operation if it was not safe.[28]

The antinuclear groups had no influence on the selection process. The government chose the investigative team. The eleven members included Manuel López Rodríguez and several Spanish engineers with experience at Spain's Lemoniz and Vandellos nuclear power plants.[29] The antinuclear activists argued that the team would not be impartial because all of the team members were strong proponents of nuclear energy. López Rodríguez was especially undesirable, given his connection to Hidroeléctrica Española, S.A., one of the companies involved in the construction of Laguna Verde.[30] He was also a friend of Juan Eibenschutz.[31] The team (which called itself the Equipo Xalapa) took eleven days to do its work, at a cost of $120,000.[32] The independent study turned out to be superficial; it was not the thorough inspection the antinuclear activists had wanted.

Once the investigation was completed, in mid-August 1990, the final report was released quickly, though access to it was restricted. The press was allowed to read the report but was not allowed to have copies. The document began by explaining that "given the number of audits and inspections that have already been done, there was no reason to undertake yet another one."[33] It indicated that the team visited the plant just once, interviewed personnel of particular sections of the plant, and reviewed documents pertaining to the plant's operation. Many of details of the team's investigation were not made public, and, except for the team's leader, not much was known about the other engineers involved in the study.[34] The team concluded that they were "conscious that our work is not complete, and there will be issues that will escape us."[35] However, their recommendation was that the plant should be allowed to operate commercially.[36]

The antinuclear activists were outraged. Pedro Lizárraga of CONCLAVE declared that the investigation had been a farce and that the Mexican government was succumbing to various national and international pressures: "The obstinacy of operating a nuclear plant that is eighteen years old, obsolete, dangerous, with two acknowledged radioactive leaks and with exceedingly high production costs and unrecoverable investments, shows that the investigative team has responded to financial and industrial interests, technical sectors of power, and definitely to strong global economic interests."[37]

At a subsequent meeting with President Salinas, the Madres Veracruzanas complained about the message in the leaked fax and told him that they also believed the inspection was a farce. They showed him a copy of the fax, and he responded by saying that he would handle it, but nothing further happened. Sara González of the Madres Veracruzanas vowed to continue the struggle: "With the operation of Laguna Verde, the anti-nuclear groups suffered a profound disappointment; the struggle did not have an effect, but now [the groups] are prepared to fight more aggressively because we have a time bomb a few kilometers from our homes."[38] Some of the activists were pessimistic, however. Thomas Berlin, author of a book condemning the plant, declared that "only an accident will stop the project."[39]

To add insult to injury, not only did the independent audit allow the operation of Laguna Verde to proceed unimpeded, but only a few weeks later an announcement was made concerning Unit II of the plant. Up to this point, Laguna Verde had been operating on only one of its two units, which went on-line in October 1988 and was working at 80 percent capacity. The second was yet to be completed. The head of the plant's Center for Information, Vinicio Serment, announced that the federal government had earmarked an additional $350 million to enable Unit II to begin commercial operation in two years, in 1993.[40] He declared that the $350 million would be used for salaries, training, and technical materials. He also defended the results of the independent study: "The results are reliable, despite the ecologists, who are never content."[41]

Although the Madres Veracruzanas were successful in securing meetings with President Salinas and other important officials such as the governor of Veracruz—access that no other antinuclear organization had had—and although they had extracted an important concession from President Salinas—the promise of an independent investigation of the plant—they, along with the rest of the groups in the movement, were disappointed with the results. They were convinced that the investigation had been a sham, and they promised to continue to work toward closing the Laguna Verde plant.

Repression

Throughout the late 1980s and early 1990s some of the most important movement leaders experienced harassment and threats. José Arias Chávez said that he had received numerous telephone calls in which threats were made on his life. Moreover, in an attempt on his life, his house in the state of Mexico was burned to the ground.[42] The house, built in 1967, was the first environmentally sound, self-sufficient structure in Mexico. Arias Chávez also said that other movement leaders had experienced similar threats and harassment. The poet and activist Luis Barquera received death threats over the telephone,[43] and Pedro Lizárraga, a university professor from Xalapa, was beaten by the police and experienced other harassments, such as finding pictures of his family, taken by an unknown person, at his desk at the Universidad Veracruzana.[44]

On April 6, 1989, however, a more serious incident occurred: Juan Marín of CONCLAVE was shot by an unknown gunman.[45] He recovered from his wounds, but no arrests for the attack were ever made. Antinuclear activists were convinced that Marín had been shot in order to intimidate the entire movement, and they organized a protest to demand that the authorities investigate the incident. The Madres Veracruzanas, however, did not experience this level of repression, despite their visibility. Every Saturday, as they protested in front of the Governor's Palace, a police officer watched and took notes, but no further action ever resulted. The Madres were never physically attacked and did not receive threatening phone calls. Being upper-middle-class and female may have protected the Madres from the harassment other groups experienced.[46]

Political Ramifications of the Laguna Verde Project

While some movement participants were analyzing the technical aspects of the nuclear problem, the Grupo de los Cien and other groups turned their attention to the political ramifications of the nuclear project. Homero Aridjis of the Grupo de los Cien argued that in his analysis, nuclear technology, because of all its inherent risks and consequent need for security, brings with it increased political repression and surveillance.[47] He came to this conclusion after studying nuclear technology's effects on states and societies throughout the world: "Most nuclear plants are converted into military zones, where the movements of the population are systematically restricted. In the name of the plant's security, soldiers receive orders to repress those who might threaten the plant. In this way, nuclear plants slowly become symbols of military repression."[48]

For Aridjis, the connection between repression and nuclear technology meant that the local population of Laguna Verde had to live with a military presence. It governed their daily lives and added to their already heightened fears about the possibility of a nuclear accident. To make matters worse, the CFE was discussing the possibility of building more plants, which would mean that this pattern of military presence and repression would be repeated in other areas of the country. An additional problem would be paying for these plants, given the expense of nuclear technology. Aridjis concluded that Mexico needed to hold a national referendum on nuclear technology in order to gauge the public's opinion.[49]

Other antinuclear groups were similarly engaged in discussions concerning Laguna Verde and citizens' rights under the Mexican Constitution. In an open letter to the Mexican people published in several newspapers, 135 antinuclear groups—including the Madres Veracruzanas, the Grupo de los Cien, and the Pacto de Grupos Ecologistas—declared that many articles of the Constitution had been violated in the government's effort to operate the plant: Article 4 (the right to good health), Article 14 (the right to life), Article 14h (the right of the people to exercise their sovereign will), and Article 39j (the right of citizens not to be attacked or intimidated by the military or navy).[50] The activists pointed out, moreover, that various parts of the Veracruz state penal code had also been violated. Article 211 specifically calls for the punishment of anyone who damages the atmosphere or who pollutes, degrades, or poisons the land or water in the state. The groups demanded that the Constitution be respected and that an independent and thorough study be performed on Laguna Verde. Additionally, they demanded that the military be withdrawn from Palma Sola and that a referendum be held on the nuclear issue.[51]

*   *   *

The Madres Veracruzanas, and only the Madres, were granted a meeting with President Salinas. The president promised them that an independent body would study the plant and if problems were found, Laguna Verde would be shut down. The Madres argued, however, that the subsequent investigation was rigged—conducted by pro-nuclear engineers handpicked by the CFE. Indeed, it appeared to be so, because after a very short visit to the plant, the technicians declared that Laguna Verde had no problems and should continue to operate. The movement was unable to convince the government that the plant should be shut down, and President Salinas did not seem interested in furthering the democratic process. Aside from the Madres, he refused to meet with any other antinuclear groups.

During the year and a half after the reactor was loaded the movement also experienced significant problems. The cattlemen's leader left the movement and joined the PRI. The Madres Veracruzanas also expelled their unofficial leader, Rebeka Dyer, and fought among themselves when faced with sending a limited number of representatives to meet with President Salinas. These problems arose because of the structure of their organization: in typical New Social Movement fashion, the organization does not have a formal structure but instead relies on spontaneity and consensus when decisions are to be made. The Madres overcame these problems, but the fluidity of the group's structure was at least partially responsible for these crises.

8 The Decline of the Movement

Only the Madres Veracruzanas Remain

By 1991, many of the antinuclear groups had ceased to exist. Almost three years had passed since Laguna Verde had gone on-line, and many of the movement participants felt disillusioned by the government's indifference to their arguments. Only the Madres Veracruzanas remained as active as before: they continued to hold their Saturday protests in Xalapa's main square and insisted on meeting with more government officials. Other groups—including CONCLAVE—could count on only a few members' participation.

During the years 1991 through 1995, those antinuclear activists still participating grew more cynical. They no longer believed that Laguna Verde would be shut down in the near future and realized that their struggle would take years; nevertheless, they continued to expose the plant's flaws and question the government's credibility. In addition, many of the participants blamed the movement's decline on their leaders' defections for positions in the government and away from their antinuclear activities. This period of movement activity was difficult for the antinuclear movement. For the Madres Veracruzanas especially, it was characterized by a desperate search for a new, successful strategy to persuade the government to close the plant.

The Madres Meet with Miguel Alemán Velasco

Though the Madres had been unable to get the government to respond to their concerns and complaints, they nevertheless continued with their plan to attempt to convince high-level officials of the severity of Laguna Verde's problems. By early 1991 the population of the state of Veracruz was focused on an impending election for the state's federal senator. Many of the Madres believed the front-runner in the race to be Miguel Alemán Velasco, a wealthy businessman and son of former president Miguel Alemán Valdés. As part of his campaign, Alemán Velasco had begun touring the state to drum up support for his possible candidacy on the PRI's ticket. During one of these campaign trips, in July 1991, Alemán Velasco swam in the Gulf of Mexico in the area of Laguna Verde to convince the populace that the water was not contaminated and that the nuclear plant was safe. Many of the state's residents, however, could not believe that he had actually ventured into the water around Laguna Verde, given the antinuclear groups' many reports that Laguna Verde was contaminating the Gulf waters.

The Madres Veracruzanas decided that they had to meet with Alemán Velasco: they could not let his deed go unchallenged, and they also believed that he was likely to be Veracruz's senator and would therefore soon be in a position to influence policy concerning nuclear energy. The group asked one of its members, Sara González, to contact Alemán Velascos's office. Because González's family knew him, the group was confident that she could schedule a meeting with the candidate.

The Madres met with Alemán Velasco on July 27, 1991, at the Hotel Xalapa; the meeting was videotaped and covered by local newspapers. Right away, at the beginning of the meeting, the Madres asked Alemán Velasco to explain his position on Laguna Verde and on nuclear energy in general. Initially, he responded tentatively, saying that "he did not have the necessary information to give an opinion."[1] Although he had made Laguna Verde a campaign issue, he now didn't seem to want to discuss the topic. The Madres refused to let him off the hook, replying that they could not believe that such a well-educated person, who had always kept abreast of the important issues in Veracruz, could be unaware of the basic points in the Laguna Verde debate.

Next, the Madres tackled Alemán Velasco's earlier comment to the press that Laguna Verde was the second safest nuclear plant in the world. They challenged him to provide the source of his information and countered with their own information, which indicated that in Brazil and England, Laguna Verde was considered one of the riskiest nuclear plants in the world.

In addition, the Madres challenged the charge Alemán Velasco made during his campaign that the opponents of the plant lacked basic information about nuclear energy. They explained that they had spent four years gathering and analyzing information from local, national, and international sources, and had personally visited and inspected the plant. They were especially disturbed by his announcement that the waters around Laguna Verde were perfectly safe.[2] They were convinced that the plant had contaminated the water in the Gulf of Mexico, and they demanded that he execute a thorough epidemiological study of the waters if he indeed won the election.

One of the Madres also touched on a sensitive political issue. When Guillermina Domínguez said that she had been a loyal member of the PRI for many years but now found herself at odds with her party on the nuclear energy issue, Alemán Velasco responded brusquely, "Well, vote for the opposition; perhaps in that way you'll resolve your problem." She said, "I'm thinking seriously of doing just that."[3]

Alemán Velasco's demeanor during the meeting ranged from authoritarian to condescending. After his brusque exchange with Guillermina Hernández, he patronized the Madres, advising them not to worry too much about Laguna Verde. After all, "we are monitored by the UN, the World Nuclear Commission, by Japan and Europe."[4] The Madres responded by saying that through Radio America, a U.S. radio network, they had heard of an accident at Laguna Verde that had gone unreported in Mexico. Alemán Velasco was skeptical, saying that he had never heard of such a radio network and cautioning them, "Don't believe it because if an accident occurred at Laguna Verde, we would immediately have the North Americans here."[5]

The Madres and Alemán Velasco were able to agree on one point—the fact that more information on Laguna Verde was needed in Veracruz. He promised that regardless of whether he won the election, he would bring to Xalapa internationally known nuclear scientists and the founder of the plant so that they could engage in a dialogue with the antinuclear activists.

The Madres were not able to convince Alemán Velasco that the Laguna Verde plant was flawed and should be closed. Nevertheless, their dialogue continued for a period of time as he provided them with books and other sources of information to support his point of view. His debate with the

Madres was relatively well publicized, so they felt that the dialogue had helped to keep the issue fresh in the public's mind.

Alemán Velasco's insensitive and often condescending treatment of the Madres provoked a backlash against the politician among many residents of Veracruz, however. In an editorial, well-known journalist Rubén Pabello Acosta lambasted Alemán Velasco for his behavior—for swimming in the Gulf, belittling the Veracruz population's fears, and patronizing the Madres.

> In that meeting you were disparaging of the group of mothers . . . when they explained their concerns to you about the plant. . . . You told them if they did not agree with your position, they could "go to the opposition."
>
> Although one of the Madres may have suggested she would change her political party, you should have tried to bring her around, or at least responded tactfully. . . . Not only did you commit a civic and political error, but you also committed a human error because any affront to that group hurts all of us residents of Veracruz due to the fact that all of us have a mother, and the vast majority of those mothers are from Veracruz. . . . And one last line: the Madres Veracruzanas represent honesty and civic valor in the face of oppressive political brutality.[6]

Pabello Acosta's editorial reveals a great deal about Veracruz society's view of the Madres. Many citizens had a great deal of respect for the group. The fact that the Madres kept their distance from politics won them much support and respect.

Despite Pabello Acosta's sharp criticism, Alemán Velasco went on to win the election and became Veracruz's senator. Though the Madres opposed his candidacy, they believed that the public meeting with him had allowed them to keep the nuclear energy issue alive before the public of Veracruz—a significant step, given that the movement had lost so many members and the era of public demonstrations had ended.

After the meeting with Alemán Velasco, the Madres demanded that the government start providing more information about the plant's operation. Initially, CFE officials had held weekly news conferences to keep the public informed, but, the Madres criticized, they had not done so for the past year and a half.[7] The women indicated that they had been relying on information provided anonymously by workers at the plant, but recently these workers had stopped supplying the information because their superiors had threatened to fire them. The workers knew that these threats were not idle because they remembered the Valdovinos Terán case from the recent past.

The Madres also focused their attention on the government's interpretation of the environment, human rights, and the Mexican Constitution. In June 1992, during a forum on human rights and the environment sponsored by the state of Veracruz, a member of the Comisión de Derechos Humanos (part of the Veracruz state government) had declared that the concept of harm to the environment and to public health stood outside of the Mexican legal system. A Mexican court would hear a case involving this issue only if an individual had actually been physically harmed. In other words, the Mexican legal system did not recognize or define potential harm as a legal problem.[8] The Madres responded angrily to this announcement. They asked, "Is it necessary that our health be harmed or that we die before our rights are considered?" Also, "When will our national political structures aimed at 'protecting the environment' cease being a farce and truly have a moral responsibility toward the community?"[9] They denounced the fact that the participants could not speak freely at the forum. Expecting or hoping for thorough answers to their questions about the Laguna Verde problem, they received only incomplete and evasive replies from authorities and other participants.

The *Proceso* Article

The antinuclear groups were relatively quiet until January of 1993, when *Proceso,* a highly respected weekly political magazine, published an article accusing the Laguna Verde plant of having very serious problems.[10] The article, written by Guillermo Zamora and based on anonymous sources working at the plant, revealed that the plant was contaminating the atmosphere and the surrounding Gulf waters, and that rust was a serious problem in many of the plant's systems. Zamora's informant also provided him with internal documents that supplied the evidence for the accusations: the "Notificación de fuentes reportables," which described events that put the plant's operation at risk; the evaluation by the Comisión Nacional de Seguridad Nuclear y Salvaguardas (CNSNS); descriptions of the causes of the plant's many stoppages; the annual report of monitored radioactive levels at the plant; and reports concerning rust and radioactive leaks into the atmosphere.

The anonymous informant said that in September 1992 Laguna Verde had been very close to suffering a major accident. At that time, the plant experienced a complete loss of external sources of electricity. Luckily, the plant had not been operating; otherwise, "all of the security systems would have been disabled, and an accident would probably have occurred."[11]

The source explained that the plant was not operating efficiently. The condenser was rusting and corroded, but the engineers continued to operate the plant instead of addressing the problem. As a result, radioactive waste had been released into the atmosphere and into Gulf waters on more than one occasion. Plus, the plant's cooling system was not operating correctly, meaning that water used in the cooling process was being exposed to carcinogens such as strontium 90, cobalt 60, cobalt 58, iodide 131, and cesium 137. All of these carcinogens were then dumped into the Gulf so that fish and other sea life were being contaminated.[12] Zamora argued that it was difficult to see how the plant could possibly operate for forty more years because in just three years it showed high levels of deterioration.

Zamora also investigated the plant's evacuation procedure—the PERE—and found that the infrastructure supporting it was practically nonexistent. There were very few shelters for the surrounding population, and most of them had no running water or medication for people suffering from radiation sickness. At the port of Veracruz, the Secretaría de Salud (Ministry of Health) was supposed to have been running a laboratory to monitor levels of radioactivity in water and food, but the laboratory did not exist. Personnel at the Hospital Regional, run by the Ministry of Health, said that the laboratory would be installed at a future date. "No one could answer why the laboratory was yet to be built when the Secretaría de Salud had provided Dr. Cristina Nava, the head of Occupational and Environmental Health, with 2.5 billion [old] pesos in 1989 to do just that."[13]

The Madres Veracruzanas and CONCLAVE welcomed the *Proceso* article. Mirna Benítez, one of the Madres, argued that part of the problem was that institutions such as the CNSNS, which oversaw the plant's security systems, were both promoters of the plant and its judges.[14] In other words, the institutions charged with the responsibility of overseeing the plant were not objective, and therefore it was impossible to believe the pronouncements of the CFE that there was nothing wrong with the plant. Benítez also discussed the issue raised by the fact that the sources at the plant were anonymous: "It is obvious that you cannot cite the source because of fear of reprisals; the workers never give you their names. A while ago they [plant officials] fired the workers associated with internal dissidence; they have reinforced their security measures so that information won't be leaked, but even now [certain] information can be obtained. There was a time when we had access to those reports [internal documents], but now that is not possible."[15]

Benítez concluded that there was only one solution to the problem: Laguna Verde needed to undergo a true, unbiased inspection. The Madres said that they had asked for a meeting with Governor Patricio Chirinos to discuss the possibility of such an inspection, but the governor had not yet agreed to meet with them.[16]

Whereas the Madres Veracruzanas believed that meeting with the governor and other officials could help them meet their goal, Pedro Lizárraga of CONCLAVE was more cynical: "We have nothing more to say to the government; we have spoken to them every Saturday for the last five years, and they have not listened. The evidence is there, and they should assume responsibility; radioactive waste is being dumped in the ocean, and radiation is affecting the people of Veracruz. Not to close the plant would be criminal."[17]

The government's reply was predictable. Interviewed during a meeting with the head of the state's fisheries department, Governor Patricio Chirinos denied that there were any problems at Laguna Verde. He had been in touch with officials at the CFE and CNSNS, and they had denied the allegations. He also told reporters that the *Proceso* article did not cite sources, and therefore it was difficult to believe the accusations. Further, in response to the Madres' complaints about the government's hesitation to give the public any information about Laguna Verde, the governor said, "I understand that at Laguna Verde there is a permanent information module where anyone can get information."[18]

Shortly thereafter, Laguna Verde officials put out their own statements. They invited reporters to the plant to persuade them that all was well with the plant. They said that the plant was completely safe. Yes, some of the equipment indeed had corrosion problems, but they were completely normal, given the materials that the plant processes.[19] They reassured the public that the rust situation was routinely cleared up when it appeared. Further, although admitting that some of the condenser's pipes did have fissures, they explained that these also were normal and that they did not affect the plant's operation. Finally, the officials denied that the plant was contaminating the atmosphere and ocean: the monitoring system indicated that the plant had not affected the environment or the population of Veracruz. Hector Luna Lastra, director of the Centro de Información Energética of the CFE, dismissed the possibility of an accident occurring at Laguna Verde: "This plant was made to withstand idiots [pendejos]; in the event that an erroneous command were to be given, due to human error, the reactor would stop automatically, and there would be no accident."[20]

The Madres continued to meet with government officials, although they had no success with this strategy. On March 16, 1993, they met with Salvador Mikel Rivera, a subsecretario de gobierno, third in command in the state's government.[21] The Madres hoped that the new governor, Patricio Chirinos, would be more sympathetic to their cause, so they presented evidence to Mikel Rivera and Francisco Morosini, head of Asuntos Ecológicos (Environmental Affairs), indicating that the plant was not functioning properly. Mikel Rivera listened carefully to their arguments and offered to form a commission made up of government officials, antinuclear activists, and pro- and antinuclear engineers. He proposed that this commission work closely with the Secretaría de Salud and that it study the plant and engage in epidemiological studies in the region surrounding Laguna Verde in order to ascertain whether the plant was malfunctioning and contaminating the environment. The Madres were promised that they could nominate engineers with antinuclear views so that all sides could be represented on the commission. Although the commission and its proposed activities sounded promising to the Madres, they were extremely cautious in their reactions:

> No mention was made of closing the plant, but we did find a certain accessibility [on their part] in listening to our concerns. . . . We are going to participate [in the activities of the commission] because we encountered a certain accessibility in the present administration. But though we are pleased in certain ways, we have our doubts because we have already experienced certain disappointments. Now we are hopeful that something will be done. We will do everything possible on our part so that what is done is done well.[22]

Indeed, the Madres heard very little about the new commission for the next few months. Though their main strategy throughout this period was to attempt to meet with government officials, they had never given up on other activities, such as attempting to forge links with environmental and antinuclear organizations around the world. Up to this point they had not been very successful in fomenting support internationally, but their luck changed in November 1993, when they received a letter from an anti-nuclear group in Germany—Energie und Umweltzentrum—in response to the information they had sent to antinuclear groups throughout the world. In a letter addressed to President Salinas, the German activists encouraged him to order an impartial investigation of the plant. "We know that the company MHB Technical Associates of San Jose, California, can execute the study."[23] They wrote that Laguna Verde was an international problem "not only because it can affect other countries, but because your country is part of the world," and they encouraged Salinas to consider the interests and concerns of the citizens who had elected him to office. In addition, they argued that Mexico was blessed with a climate ideally suited for the use of alternative energy sources, such as solar energy.[24]

A few days later, President Salinas received a similar letter from the Belgian deputy Marguerite-Marie Dinguirard. She encouraged Salinas to follow the principles of sustainable development and added that the European community knew about the existence of Laguna Verde, "whose operation evokes much concern in the local as well as international community."[25] President Salinas never responded.

Mirna Benítez Runs for Office

During 1994 the antinuclear activists turned their attention to the presidential succession. Cuauhtémoc Cárdenas of the Partido de la Revolución Democrática (PRD) once again announced that he would run for the presidency against Ernesto Zedillo of the PRI (who ran when Luis Donaldo Colosio was assassinated in March 1994) and Diego Fernández de Cevallos of the PAN. Cárdenas reiterated his position against Laguna Verde, promising that he would close the plant if he were elected. At this point, after seven disappointing years of participating in the antinuclear movement, some of the Madres decided to change one of the most fundamental principles of their struggle—keeping their distance from political parties.

Cuauhtémoc Cárdenas approached the Madres, asking if one of their members wished to run for the Senate as a PRD candidate. After much deliberation, Mirna Benítez decided to accept his offer. The Madres were divided on the issue. Some continued to argue that the group should remain steadfast in keeping political parties at bay. Others maintained that they had to take advantage of this opportunity ("aprovechar la coyuntura") because it might lead to further antinuclear mobilization. The decision was not an easy one for the Madres. Because there was no consensus, Benítez decided she would withdraw from the group during her campaign.[26] In addition, the Madres allowed Benítez to run only after getting reassurances from the PRD that the group would keep its autonomy: they were by no means joining the party en masse. In fact, the ideologies of individual members of the Madres Veracruzanas covered the entire Mexican ideological spectrum; privately, they belonged to all three political parties, and they intended to stick with their respective parties.

Pedro Lizárraga of CONCLAVE believed that, for the Madres, this new strategy of political affiliation could have positive and negative aspects.[27] On the one hand, joining forces with the PRD would spread the anti–Laguna Verde message further, which was important because as time had passed, the antinuclear issue had lost much of its urgency for the general population. Yet if one of their members ran under the PRD banner, the Madres risked alienating certain political factions—especially the PRI, the dominant force in Mexican politics. According to Lizárraga,

> In Mexico, where the entire administrative apparatus is controlled by the government, that is to say the PRI, surely the new nexus of the Madres Veracruzanas will close even more doors for them regarding their demands for an impartial technical audit and regarding possible conversations and negotiations. But if the candidacy of Mirna Benítez ends the amiable treatment [that they have received], it also ends the governmental hypocrisy; the Madres have never obtained anything substantial.[28]

Mirna Benítez, like Cuauhtémoc Cárdenas, lost the election, so the Madres had to examine future strategies for participation. Benítez believes that her bid for the Senate helped the movement, however. Throughout the campaign, crowds gathered at the Plaza Lerdo once again, and many of these people carried antinuclear banners. Laguna Verde was once

again an important topic of conversation. However, after the election—
held on August 21, 1994—the mobilization came to a halt.

After Benítez's unsuccessful campaign for the Senate, the Madres continued to ask for meetings with government officials. In October 1994,
six and a half years after beginning their weekly protests, they announced
that they were seeking a meeting with Governor Patricio Chirinos. During an interview at their weekly Saturday protest, Mirna Benítez acknowledged that the Madres were having difficulty acquiring information about
what was going on at Laguna Verde.[29] A few newspapers had published
articles indicating that plant officials were building a laundry facility to
wash contaminated uniforms. Benítez said that the Madres worried that
water from the laundry would harm the environment: "If the laundry is
to be built at El Farallón, municipio of Alto Lucero, information should
be provided about where the contaminated water, affected by the washing
of the uniforms, will be dumped and whether this water will undergo
treatment before it is allowed to spill out. If this water does not undergo
treatment, then the lagoon at El Farallón will run the risk of being contaminated by radioactivity, which in turn will kill off the area's flora and
fauna." Benítez and the Madres were especially worried about fishing
activity in the area. Fishermen had reported discovering several fish kills,
and the Madres worried about consumers eating contaminated fish caught
in the Laguna Verde area.

Benítez and the Madres admitted that their struggle against Laguna
Verde had become extremely difficult. For several months they had contact with only lower-level government officials, and their sources of information at the plant had dried up. They desperately wanted a meeting with
Governor Chirinos: "We don't want a meeting with the subsecretary of
government; now we want an audience with the governor so that he can
explain to us what is happening inside the plant. As the former secretary
of urban development and the environment [the head of SEDUE], he should
know the conditions under which the plant operates."[30]

Lack of information about the plant, as well as government noncoop-
eration, was causing the movement to lose participants rapidly. Benítez
and the Madres admitted, "[We are at] a very difficult stage in our
struggle."[31]

More Leaders Defect

Between 1993 and 1996 the Mexican antinuclear power movement lost the majority of its early participants: in those years practically the entire antinuclear movement consisted of activities conducted by the Madres Veracruzanas. The decline in numbers is not difficult to understand. Regular members of the Cattlemen's Association, as well as antinuclear participants in general, lamented that several prominent leaders had been co-opted by the government, and they blamed this process for the movement's decline. The cattlemen from Palma Sola no longer voiced their opinions about the plant. Not only did they fear further repression by the government but they also felt that their leaders had betrayed them. The five hundred members in the area often recalled that their former leader, Efrén López Meza, had left the movement and in subsequent years had served as a local deputy and later as mayor of the city of Veracruz. The cattlemen said, "Many environmentalists used the movement for their own purposes and have forgotten about us."[32]

Other movement participants pointed to the alleged co-optation of Fernando Jácome (former head of an antinuclear group from Coatepec, Veracruz) and of Efrén López Meza. Antonio Viveros Salas of the village of Emilio Carranza (near Palma Sola) discussed the co-optation of leaders and how it had affected the antinuclear movement. One of the earliest participants in the movement in the Palma Sola region, Salas believes that the cattlemen's leaders left the movement because "they convinced them"[33] to do so; that is, the government persuaded the leaders through political positions and monetary rewards. When the leaders accepted these offers, the alliance with the local population was ended, and the government disarmed the movement ("los aplacó").[34] Political positions were especially important in the case of López Meza and Silvestre Morales, a former cattleman who became mayor of the town of Vega de Alatorre, Veracruz. Although movement activity has come to a standstill in the area, Viveros Salas insists that the population is still firmly against the plant. "We have to keep fighting until we cancel the project. The population is in constant fear because of the danger posed by the contamination. I shall keep on fighting, and I'll be an antinuclear activist all of my life because I know the dangers posed by radioactive contamination."[35]

A female antinuclear participant from the area had similar opinions. Rosa María Palafox Rendón from the town of Emilio Carranza was hopeful that the local population could be mobilized once again, but she was also bitter about the fact that certain leaders left the movement. She was especially disappointed in Edith Morales Aguirre, who left the antinuclear movement for a position with the municipal government of Vega de Alatorre, Veracruz.[36]

•   •   •

The Madres Veracruzanas experienced defections, and they expelled one of their most active and dynamic members. Yet these events did not lead to the organization's demise. Certainly the group was not as large as it had been a few years earlier, but it continued to hold its weekly protests on Saturdays and to challenge the government's position on Laguna Verde at a time when no other antinuclear group was still functioning.

Why was this the case? Why did the Madres continue when the rest of movement had collapsed? Here, the New Social Movements literature is especially insightful. The Madres, more than any other Mexican antinuclear group, displayed Alberto Melucci's concept of the "submerged network,"[37] that is, they built affective ties that linked their everyday lives together. The Saturday protests, along with the weekly Wednesday meetings, provided a routine that existed in no other group. This steady contact, in turn, produced intense friendships that transcended movement activities. Although the Madres' organization (in classic New Social Movement fashion) is officially nonhierarchical, in reality the group has been directed by two strong leaders throughout the years of protest. In the beginning, Rebeka Dyer was an effective and dynamic leader, but since her separation from the group, Mirna Benítez has helped the Madres to preserve the group's focus. In addition, the Madres never experienced state repression. While other organizations were devastated by beatings, threatening phone calls, and a military presence (in Palma Sola), the Madres Veracruzanas were left alone. Consequently, this group was the only one still participating in the antinuclear movement in early 1996.

9 Mothers' Movements and Feminist Theory

The Case of the Madres Veracruzanas

Are the Madres Veracruzanas alone in choosing to mobilize based on an identity of motherhood? Certainly not. Numerous mothers' organizations have emerged recently in various contexts throughout the world. Although perhaps the most famous of these mothers' groups is the human rights organization the Madres de la Plaza de Mayo of Argentina, others pursuing various goals have sprung up throughout Latin America and beyond. The members of these mothers' groups argue that they have extended their roles as protectors of children beyond the private domain of the household into the public and political arenas. The Madres Veracruzanas have specifically maintained that they want to protect their children from the dangers of nuclear technology by demanding that the Mexican government close down the Laguna Verde nuclear power plant off the coast of the port of Veracruz.

Because mothers' organizations seem to be increasingly common, feminist scholars have begun to debate the significance of such groups. Some feminist theorists argue that mobilization based on an identity of motherhood is a dead-end strategy because it merely reinforces old stereotypes of sex role differentiation. Other scholars have challenged this assumption, arguing that the use of maternal imagery need not be inherently nonprogressive. My argument is that although mothers' mobilization may seem traditional and not "progressive," the Madres Veracruzanas' own movement has evolved through time, and the group has actually challenged old gender and public-private boundaries. In addition, feminist theorists should consider race and class differences among the types of mothers' groups: some mothers' groups have been amenable to incorporating women/mothers of different classes, but in Mexico the Madres

Veracruzanas have remained an upper-middle-class group, often actively protecting their class interests even as they struggled to protect the Mexican environment. Mothers' groups thus should not be romanticized; many of these groups are divided along race and class lines, as are other organizations and movements. In the following section I provide an overview of feminist theory on mothers' movements, then proceed to a specific analysis of the Madres Veracruzanas.

Mothers' Movements and Feminist Theory

What is the feminist theory perspective on the political mobilization of women based on their identities as mothers? Scholars' opinions vary, echoing several points in the more general debate about gender equality and difference.[1] Some feminist scholars have deep reservations about mothers' organizations. Micaela di Leonardo, writing on mothers' organizations and militarism, has identified several problematic tendencies in mothers' movements.[2] Mothers' organizations often make essentialist arguments about men and women based on differing reproductive functions: men are inherently warlike, whereas women are "naturally" peace loving. This distinction is part and parcel of "Moral Mother" imagery in which women are viewed as inherently more protective of life.[3]

Such maternalist imagery is not new. "Through the manipulation of images of women as morally superior mothers and wives," di Leonardo contends, "nineteenth and early twentieth century feminists claimed the right to enter the public world as moral reformers and 'social housekeepers.'"[4] These movements' participants thus maintained that they were preserving the sexual division of labor: they entered the male, public, political sphere only in order to correct a particular problem.

Di Leonardo proposes three arguments against the use of Moral Mother imagery by female activists. First, although this imagery may be appealing at first blush, "these newly mobilized women then have no reason to become feminists."[5] Moreover, it essentially precludes a thorough analysis of military processes: "The Moral Mother argument is a poor organizing tool: it does not challenge us to think in complex ways about the sources of military threat, nor about women's own consciousness and social activity."[6] Second, di Leonardo believes that Moral Mother imagery privileges heterosexual mothers while casting aside childless women, lesbians, and antimilitarist males. Finally, such imagery is vulnerable to empirically based counterarguments: in real life, women are not necessarily peaceful or moral, and women in fact are now joining the military in record numbers.

Writing about the Madres de la Plaza de Mayo of Argentina, María del Carmen Feijoo has similar reservations.[7] The Argentinean mothers argue that their intention in the 1970s and early 1980s was to protest the disappearance of their children by extending their roles as mothers into the political arena. The Madres' actions, according to Feijoo, were an "extension of the sexual division of labor in Argentina, which gives mothers the responsibility of defending and protecting their sons and daughters."[8]

Feijoo believes that the Madres de la Plaza de Mayo served an important function in confronting the military government's human rights abuses and in developing new political practices. The mothers were admired throughout the world for their courageous confrontation of a military regime, something that no other groups in Argentina were willing or able to do. In addition, they developed novel political practices, such as the weekly protests in the main plaza of Buenos Aires and the wearing of white handkerchiefs, which came to symbolize their resistance to the regime's policies.

The Madres' political strategy was ultimately self-limiting, however, and had at least two problems, according to Feijoo.[9] First, their organizational style, which responded well to crises, subsequently was ill suited to working within the democratic institutions that were erected under President Raul Alfonsín. "The Madres as a group had a weak institution with minimal functional differentiation; they rely heavily on strong personal leadership and are held together by gender solidarity, not organizational sophistication."[10]

From a feminist perspective, the second problem with the Madres' political strategy concerns the fact that mobilization behind the banner of motherhood ultimately serves to reinforce the traditional sexual division of labor and does not challenge women's subordinate position in society. According to Feijoo, "Linking the possibility of change to feminine emotionality constitutes a paradoxical 'vicious circle.' Doing politics based on emotions . . . ends up making altruism sacred."[11] The pursuit of altruism, however, maintains women's subordination.

Other scholars, though, have countered these arguments about mothers' movements. In her study of the U.S. organization Women Strike for Peace (WSP), Amy Swerdlow defends the notion of women mobilizing as mothers. This group, which was especially active throughout the 1960s and early 1970s, challenged both the United States and the Soviet Union to end the nuclear arms race; eventually the organization also opposed the Vietnam War.[12] Swerdlow argues that it is not surprising that these women mobilized based on their identity as mothers; the participants had been raised in the 1940s and 1950s, during a time when women had been socialized to believe that motherhood was the most important aspect of a woman's identity.

The organizers of WSP used maternal imagery for at least two reasons. First, they were "expressing their own sense of male betrayal of the agreement they, as women, had made with society to sacrifice their own personal interests and career goals in favor of raising the next generation."[13] That is, because male political leaders were putting much of the world's population at risk by pursuing a nuclear arms race, it was up to women, as nurturers, to step into the public sphere to save the world. Second, these women were mobilizing as mothers because "they were also trying to speak to the American people in a language they believed would be understood by the American people" and because they also were fearful that if they mobilized as anything other than mothers, they would be dismissed or attacked because of sexism.[14] Swerdlow is careful to state that though the women of WSP mobilized as mothers, they never claimed to be more peaceful or more nurturing than men.[15]

Swerdlow provides counterarguments to feminist theorists' criticisms of mother's movements. Specifically, she states that di Leonardo's argument that Moral Mother imagery is a poor organizing tool is "historically inaccurate and leads . . . to a simplistic, undialectical view of women's movements for social transformation."[16] A careful analysis of WSP's movement led Swerdlow to conclude that despite the subordinating bent of maternal imagery, the women experienced a "sense of personal empowerment"[17] as a result of being part of the movement. In fact, after achieving their goals, the women did not return to the home because, ironically, they no longer regarded themselves primarily as housewives. At the end of her analysis, Swerdlow says that "maternal rhetoric can be an inspirational organizing tool, a source of energy, commitment, and passion for traditional middle-class white women."[18]

In sum, feminist theorists have identified several reasons for questioning the utility of maternal imagery in women's mobilizations. First, such maternal imagery reinforces rather than questions the traditional sexual division of labor in society. Second, it promotes an essentialist view of men and women. Some feminists also warn that maternalist mobilization can invite "counterconstruction" by governments and other elites, which can, if successful, in effect drive women off the political stage. Finally, maternalist mobilization can work well during the initial stage of mobilization, as it did in the case of the Madres de la Plaza de Mayo, but can be noticeably ineffective at working with democratic institutions in the long run.

Ultimately, the debate about mothers' movements reflects the current debate about equality and difference in feminist theory. During the late 1960s and throughout the 1970s, feminist theorists were primarily interested in exploring women's unequal status in society.[19] Theorists influenced by materialist approaches attributed women's subordination to their simultaneous oppression by capitalism and patriarchy. They thus viewed the family and private sphere as an important site of the oppression of women. Subsequently, some feminist theorists began to change their perspective, giving gender the most important (if not the only) position in their analysis. Scholars such as Sara Ruddick began to study and espouse "maternalist thinking."[20]

Those scholars who explore "difference" have moved gender to the center of analysis; they do not treat gender as a derivative category. Instead, they have begun to embrace those traditionally undervalued female traits usually associated with mothering. Ruddick bases her argument not on the biological underpinnings of motherhood, but on the work mothers actually do. That is, "maternal thinking" emerges not from hormones but from caring for and nurturing a human being, which gives mothers (who can be male if they also perform this work) a different political agenda.

In the next section, I examine the case of the Madres Veracruzanas in light of this theoretical debate between equality theorists and difference theorists.

Feminist Theory and the Madres Veracruzanas

How do feminist theorists' criticisms of mothers' movements fare in the case of the Madres Veracruzanas? Have the Madres reinforced the Mexican sexual division of labor, and have they made essentialist arguments? Have they been vulnerable to counterconstructions of the term *mother?* And have they shown an inability to work in the long term with Mexican political institutions?

First, the Madres Veracruzanas *have* worked within the traditional paradigm of sex role differentiation. They believe that in Mexican society, women are ultimately responsible for the care and nurturing of infants and children. They have been clear that their motivation for joining the antinuclear struggle has everything to do with their role as protectors of children; they argue that they are merely extending that role from the private into the public sphere. Indeed, the Madres often carried pictures of their children during the weekly protests in Xalapa's main square: the pictures served to remind government officials and the public at large that the Madres wished to protect their children from the dangers of nuclear technology.

Moreover, ever since the group was formed, the Madres have tended to make essentialist arguments. Most of these arguments have reflected their belief that many politicians and CFE officials have become so enamored of advancing technologies that they have forgotten about the welfare of the society as a whole and of children in particular. The Madres feel that it is up to mothers, as the nurturers of children, to remind politicians and engineers that not all new technology is desirable and may actually pose dangers for humans and the environment. The Madres' motto is "Porque amamos la vida" (Because we love life). Thus the Madres believe that their unique role as nurturers gives them a special insight into what is good for children and for society in general; sometimes they have to transcend the domestic sphere in order to perform their duties.

Although the Madres have consistently used maternal imagery in their mobilizational efforts, this imagery is not one-dimensional. Their identity as mothers not only reflects their view of the sexual division of labor but also symbolizes their disdain for Mexican politics in general and for the authoritarianism of the PRI in particular. The Madres often speak of their disapproval of authoritarianism, and they believe that politicians, as well as many individuals who are involved in politics, are interested only in achieving political power and economic gain. In their view, women *generally,* however, are less interested in achieving power over others. The Madres Veracruzanas work very hard to maintain the image that they are "above politics" and have no interest in economic gain. They believe in these ideas so strongly that the group expelled an important member in order to stay true to them.[21]

The Madres have stated repeatedly that they do not respect most politicians and that they want Mexican society and the government to understand that they will not participate in normal political practices—including bargaining, making compromises, or being co-opted. They argue that nuclear energy is so dangerous that they will back down only after Laguna Verde is closed. For them, there is no room for compromise, and they insist on disciplining group members in order to avoid co-optation by a corrupt and authoritarian political process.

Not only is maternal imagery a means to protest against authoritarianism, but it can also serve as a means for change. Although it can portray women as self-sacrificing and meek, it can also evolve in order to empower women, as it did in the case of the Madres Veracruzanas. They argue that they have changed a great deal as a result of their experiences in the movement. In the beginning, many of the women lacked confidence in the public sphere or regarded themselves as mere housewives, but now, years later, they think nothing of confronting top government officials. Moreover, they are comfortable dealing with the media: they are often asked for their views about nuclear energy and the environment in general. And most important, this sense of efficacy has carried over into their personal and professional lives. Carolina Chacón, a schoolteacher, says that she is no longer submissive in her interactions with colleagues and superiors in her professional life. She and several of the mothers say that they have "awakened"—that they now question all aspects of the status quo.[22] Thus, although initially these women may have based their mobilization on traditional sex roles, they are now in the process of negotiating the boundaries of those roles.

Another important issue in mothers' movements is the risk they run of counterconstructions of the term *mother* by political and other actors. The Madres have been very conscious of the possibility of negative counterconstructions, and to date they have been relatively successful in avoiding them. Female and indigenous activists have been subject to the most venomous attacks by pro-nuclear interests. Technology is perceived to be the domain of white and mestizo males in Mexican society, so women and indigenous peoples are perceived to be inherently incapable of understanding the technology. They are therefore ridiculed and told that only trained technicians and (certain) professional politicians should decide the fate of nuclear technology in Mexico.[23]

Nevertheless, the Madres and their defenders have been very aggressive in addressing these criticisms. After one incident in which the Madres were pushed and shoved by members of the Confederación Nacional de Organizaciones Populares (CNOP), a group affiliated with the PRI, Pedro Garcimarrero, a prominent Xalapa journalist, wrote an editorial defending the Madres. They themselves also often took out advertisements in local newspapers in order to defend themselves.

Another issue associated with counterconstructions of the term *mother* is the question of class. Most members of the Madres Veracruzanas are from middle- to upper-class backgrounds. The Madres have thus been very careful to construct the term *mother* in a particular way. For example, the Madres' political tactics have been carefully calculated in order to preserve their pure, dignified image because they want to be perceived as being above politics. The mothers also despise the tactics of the working class (including working-class women), which they say often involve the use of obscene language and (sometimes) violent behavior. They also disapprove of the tactics of feminist groups from the United States and Western Europe. The mothers also explain that the range in their ages—from approximately thirty-five to fifty-five—helps their image. It works to their advantage because the media and Mexican society in general do not view them as mere young sex objects who are participating in the movement only to attract attention to themselves. Instead, they are regarded as serious, concerned mothers who are interested in protecting their children's welfare.

The Madres Veracruzanas have attempted to define the terms *woman* and *mother* (they use the two interchangeably) so that the terms reflect not only their gender but also their own particular class backgrounds. For example, when Mercedes Solé says that Mexican women would never chain themselves to fences, she is referring only to Mexican women of the middle class. Of course, she and the other Madres Veracruzanas are aware that poor and working-class Mexican women in organizations, including antinuclear organizations, often do use such tactics.

The Madres' construction of the terms *women* and *mothers* has led to the exclusion of poor and working-class women from their group. The following example illustrates this exclusion and the class issues in the

Madres Veracruzanas' mobilizational strategies. As noted above, the Madres insisted on group discipline—a certain type of "feminine" attire, no harsh or "obscene" language, and no taking over public buildings, for example.[24] These rules precisely encompass some of the tactics that many poor and working-class women use in the antinuclear and other movements in Mexico.[25] Yet the Madres are truly concerned about recruiting more women of all backgrounds for their organization. The members have expressed a great deal of concern about the fact that the movement and their organization are not growing.[26] They believe that this could be problematic if the government begins to think that the Mexican public is no longer concerned about the nuclear debate.

The Madres refuse to change their approach on class issues, however, and often seem unaware of the concerns of poor and working-class women. During the spring of 1990, for example, the antinuclear organizations received some disturbing news from a fishermen's cooperative located near the Laguna Verde plant. The fishermen had begun an investigation as to why the number of shrimp had declined in the area, and their research pointed to nuclear waste dumped illegally in the waters surrounding the plant. Moreover, the Laguna Verde plant's analyses of water samples and shellfish from the area indicated that the shrimp had absorbed unusually high levels of strontium 90, a carcinogen. The mothers decided to exploit this information in order to mobilize more women in the fight against Laguna Verde. During their regular Saturday morning protests, they handed out leaflets in downtown Xalapa that announced the fishermen's results about the contaminated water and shrimp. The leaflets concluded by asking, "Are you willing to serve your family this shrimp during the current Lenten season?" and then encouraged women to join the movement.

Ironically, the Madres had not considered that perhaps the vast majority of Mexicans may not be terribly concerned about the contamination of a food that they are not in the habit of consuming. Shrimp is expensive in Mexico, and most Mexicans are poor. The country experienced an economic crisis through most of the 1980s; many foods consumed by Mexicans (beans, for example) skyrocketed in price. Thus, the news about a carcinogen in shrimp did not prove to be an effective tool for the mobilization of the general population, and in the end, no new women joined the organization.

The construction of concepts and terms has implications for political mobilization. The Madres' attempts to define the terms *woman* and *mother* so narrowly may perhaps have made the organization more acceptable to government officials, but these attempts have also prevented the Madres from recruiting women of poor and working-class backgrounds.

Finally, the literature on mothers' groups indicates that although they may excel at raising society's consciousness about particular issues, they are noticeably ineffective in working with established political institutions. Because of the group's informal organization, the Madres de la Plaza de Mayo, for example, have been "more prepared to respond to a crisis than to institutionalize a durable model of participation."[27]

Is this also a valid critique of the Madres Veracruzanas? I would argue that it is not. The Madres Veracruzanas, together with the larger anti-nuclear and environmental movements, have been unable to convince the Mexican government to close the Laguna Verde plant, but their failure cannot be attributed to an organizational problem. Indeed, the Madres have exhibited a high degree of sophistication in their many meetings with government officials such as President Salinas and Governor Dante Delgado Rannauro. If Laguna Verde is still operating, it is because the executive branch of the Mexican political system, which is extremely centralized, has refused to budge on this matter, despite the vehement protests of antinuclear groups.

•   •   •

Most theories that attempt to analyze and explain social movements have failed to include the significance of gender in the analysis.[28] Feminist theorists, however, have developed a literature and a debate specifically concerning mothers' movements. They have found mothers' movements to be problematic: although many of them have been very successful in awakening women to important issues and getting them involved, they also have often reinforced the traditional sexual division of labor in society while putting forth essentialist arguments. In addition, these movements often invite counterconstructions. Governments and political actors in general can argue that these mothers, if they are "true" mothers, should be home caring for their families rather than meddling in the public sphere. Finally, the Madres de la Plaza de Mayo have been criticized for their inability to engage in standard politics after democratic institutions were revived in Argentina.

The case of the Madres Veracruzanas in Mexico serves to illustrate the strengths and weaknesses of these arguments. The Madres Veracruzanas *have* reinforced the traditional sexual division of labor in Mexico. They have argued that they are in the political arena in order to fulfill their role as protectors of children and of society. They believe that the Laguna Verde nuclear power plant is dangerous for their children, and they have ventured into the political arena to convince politicians that the plant should be shut down. Yet the Madres have *also* renegotiated these gender boundaries. They believe that it is now appropriate for mothers to question the authoritarianism of the government and of technocratic elites who have made decisions about nuclear energy without consulting civil society. Moreover, the experience of meeting with important government officials and movement activists has given the Madres a new self-confidence that has spilled over into their professional and private lives. Thus, although the Madres continue to base the justification of their political participation on the Mexican sexual division of labor, this division has evolved into something new.

Feminist scholars have also argued that mothers' movements can be vulnerable to counterconstructions. In the case of the Madres Veracruzanas, certain groups have indeed attempted to smear the women. Pro-nuclear activists have been especially critical of the Madres and of indigenous antinuclear activists—claiming that they are not qualified to give opinions on issues of high technology. But the Madres have been able to keep these counterconstructions at bay by carefully monitoring their pure, dignified image. They maintain that they are above politics and have even expelled a leader they believed might compromise the group politically. Yet these tactics have also erected class barriers that prevent the group from recruiting working-class and other women. The Madres' dress, demeanor, and political tactics—reflecting their middle- to upper-class status—have distanced them from women who are not from the middle or upper classes. Clearly, mothers' organizations should not be romanticized: they, like many other groups, are divided along racial and class lines.

The case of the Madres Veracruzanas ultimately helps to answer the question of why some women would mobilize politically as mothers rather than as just citizens or as feminists. The Madres Veracruzanas argue that Mexican society is more likely to listen to women who call themselves mothers rather than feminists. Thus, they were constrained by societal forces: the women who founded the group believed that maternal imagery was most likely to be effective in their mobilizational efforts. Structural factors were also at play. One of the members says that she joined the Madres Veracruzanas because she believed that this group, more than any of the other antinuclear groups, would be sympathetic to her busy schedule as a full-time mother. She is one of the most active members, yet there are times when family demands interfere with her participation. She believes that the Madres understand these pressures better than the other organizations.

The case of the Madres Veracruzanas indicates that mothers' movements are not necessarily one-dimensional and that maternal imagery may not be static. The Madres Veracruzanas' mobilization as mothers not only serves to make a statement about the sexual division of labor but also symbolizes the members' rejection of the PRI and of standard political processes in Mexico. The Madres believe that the Mexican political system is corrupt and authoritarian, and they choose to confront it as mothers who have pure motives and are interested only in what is good for children and for society—that is, closing the Laguna Verde plant.

Finally, maternal imagery is not necessarily static. The Madres Veracruzanas have changed as a result of their participation in the group and in the larger environmental struggle. Mirna Benítez, one of the members, explains how this has happened for the women.

Each of the mothers has become politicized in her own way. Irma, the doctor, has come to realize, through her participation, that she is against aggressive technology. She is not against technology, but she is against destructive technology that the government has pursued only because it is new and modern. Irma favors technology that is not destructive to people and the environment. Margarita Castellanos lives out in the country, and she has grown to love her land and the vegetation on it. Her motivation for joining the group and her political analysis are rooted in this love for the land and the countryside. Each woman's position and analysis is thus rooted in her own personal experience. . . . [The mothers'] evolution has been steady. Initially, the mothers didn't see much beyond their own homes, then Xalapa; now they've grown to understand the political system and how it affects the nuclear industry in Mexico.[29]

Thus, although mothers' movements may appear to be purely traditional in that they reinforce the sexual division of labor, they have the potential to change and become multidimensional. The Madres Veracruzanas have not achieved their goal of closing the Laguna Verde plant after years of attempting to do so, but they have promised to keep struggling until government officials give in to their demands.

All in all, social movement theories do not provide a framework for studying gender dynamics in such movements, but feminist theory can fill this gap by providing the necessary categories to complete the analysis.

10 Conclusion

This study began with a series of questions. Specifically in the case of the Mexican antinuclear movement, can the New Social Movements paradigm, which emerged in Western Europe and the United States, be applied to the analysis of movements in Latin America? Many scholars of Latin American social movements believe that such movements have been at least partially responsible for Latin American governments' turn to democracy. Has this been the case in Mexico, however? The first part of this chapter discusses whether the Mexican antinuclear movement can be considered a New Social Movement—specifically in terms of the NSM concepts of autonomy, internal democracy, political practices, submerged networks, identity, and postmaterial goals. The second part of the chapter discusses the question of whether the Mexican antinuclear movement has furthered the process of democratization in Mexico.

New Social Movements and the Mexican Antinuclear Movement

The NSM perspective maintains that New Social Movements are characterized by a strong drive to preserve their autonomy from political parties. The question of autonomy was indeed of paramount importance to the antinuclear groups in Mexico.[1] The residents of Palma Sola and individual organizations always insisted on autonomy both from the larger movement and from the state. Although the cattlemen and villagers participated enthusiastically in the larger movement's events (trips to Mexico City and Xalapa, blockades of traffic), they were vigilant of threats to their autonomy. The cattlemen wanted to ensure that their particular concerns—protecting the cattle industry and the integrity of their town—would not be lost in the larger movement. In addition, the cattlemen and villagers exhibited a strong antipathy toward the state and traditional political channels. None of the villagers had any previous political experience, and many expressed surprise at finding themselves engaged in discussions with relatively high ranking politicians. Overall, many had negative opinions of the state, and their antipathy toward politicians and the state actually *increased* as time went by, especially after their discussion of the evacuation procedure (PERE) with government officials and when they had experienced repression by the military.

But perhaps the group that guarded its autonomy most strongly at all times was the Madres Veracruzanas. From the beginning, the Madres were very clear that they had only *one* goal—closing Laguna Verde. The women were extremely wary of having their interests and concerns diluted by participation in the larger movement. They were especially concerned that groups with poor and working-class members might want the women to support other class-based goals (which they opposed). At the same time, however, they were very enthusiastic about the larger movement. These simultaneous and contradictory positions meant that they had to spend much time and energy on discussions with other groups to ensure that the movement as a whole and their own specific interests were not harmed.

The Madres expressed much disdain for political parties and for the political process in Mexico. Their strategy was essentially to do an end run around normal political channels. The Madres insisted that their cause was just and pure, so they refused to use the usual political channels because they believed that politicians and politics are tainted. Instead, they arranged meetings with top officials and appealed to them directly to close the nuclear power plant. The Madres' concern for autonomy led them to take extreme positions at times. Rebeka Dyer was expelled from the group in 1988 because her husband accepted a government position. Moreover, when Mirna Benítez ran for the Senate in 1994, she felt compelled to leave the group because some of the Madres felt very strongly that the organization should at all costs stay away from any affiliation with political parties. Benítez was able to return to the group, but only after she lost the election. It appears, then, that the question of autonomy is applicable in the case of the Mexican antinuclear movement.

The claim that NSMs use novel political practices also rings true for the antinuclear movement in Mexico. CONCLAVE and the Madres devised creative ways to protest the government's decision to operate the plant. For example, the groups published *esquelas* (obituaries) mourning the death of nascent democracy in Mexico. Similarly, the Madres placed a coffin in Xalapa's main square to symbolize the death of the sovereignty of the people of Veracruz. They and other antinuclear activists also lit candles around the coffins and prayed over them. To express their opposition to the nuclear plant, the groups took practices that are common in Mexican culture and changed them slightly to draw attention to their cause.

The issue of whether the organizations within the antinuclear movement exhibited their own internal democracy is more problematic. The Madres and CONCLAVE were both very conscious of the issue of democracy. They recognized that while they were struggling against an authoritarian regime, they had an obligation to engage in democratic practices within their groups. They were not always successful, however. CONCLAVE met regularly and used democratic processes to select leaders. The Madres, however, refused to follow traditional models. They believed they could be more democratic by having no identifiable structure or leaders. During demonstrations, when the police asked them for their leader's name, they would always respond proudly that they had no leader. However, they did not actually lack leaders; further, their leaders did not always do what the members wanted. In the early years, Rebeka Dyer led the group, and when she was expelled, Mirna Benítez began to perform that function. As noted in chapter 7, this fluid leadership structure generally worked well for the Madres, enabling them to respond in a fresh, spontaneous way to crises. At times, however, it caused problems, especially when they were forced to select only a few members to meet with the president in 1989. There was so much tension involved in making the decision that the group seemed to be falling apart. After much discussion and negotiation, they did select their representatives, but some of the members remained dissatisfied with the selection process. In sum, many of the antinuclear groups did attempt to operate democratically but were not always successful.

The NSM concept of "submerged networks"[2] has been most evident in the case of the Madres Veracruzanas. The Madres have not always participated in the movement at the same pace since the group was formed in 1986. During the early years, from 1986 through early 1989, they worked at a frantic pace—protesting, blockading traffic, organizing the "red ribbon" campaign (in which they encouraged businesses and homes to display red ribbons as a symbol of their opposition to nuclear energy), and meeting with numerous government officials. During that time they received much publicity and were highly visible at the numerous demonstrations.

After 1989, however, the shock of having a working nuclear power plant in Veracruz wore off, and it became more difficult to mobilize thousands of people to protest the existence of Laguna Verde. Many people have criticized the Madres and other groups in the movement for not being as active in protests in the recent past. Yet the NSM concept of submerged networks helps one to see that, in fact, the Madres have not stopped working: they still hold weekly meetings every Wednesday afternoon at members' homes, and they continue to protest every Saturday morning in front of the Governor's Palace in Xalapa.

The NSM depiction of submerged networks can be accurately applied in this case. The Madres continue to meet in spite of many setbacks and even when the larger movement is not particularly active. The members say that they have developed deep friendships within the group, and they look forward to seeing their friends every week. However, their meetings are in no way mere coffee klatches; even when they are not involved in a particular crisis, they keep busy by reading nuclear energy literature from around the world. The women are constantly gathering evidence about the dangers of nuclear energy in preparation for future meetings with top government officials. One of the Madres, a linguist, translates pertinent articles and sections of books so that the members can keep abreast of information from the United States and Europe. Indeed, whenever the Madres have met with officials, such as President Salinas, they have always been prepared to state their case and have backed up their position with information packets to help convince the officials of the severity of the dangers of nuclear energy.

Further, the submerged network has served another useful function. The vast majority of the Madres Veracruzanas belong to the middle and upper-middle class. Many of the members have full- or part-time servants, which allows them to spend time working against nuclear energy. Nevertheless, at times individual members have experienced problems with their husbands, who have become upset about the amount of time that they devote to the movement. Each time this has happened, fellow members have been extremely sympathetic and have provided a member with concrete suggestions to ease the problem with her husband and still remain in the movement. Thus, the submerged network works in the sense that friendships help not only to cement the group together but also to maintain members' relationships with their own families, which in turn helps to ensure continued participation.

The NSM question of whether individual participants experience a change in identity[3] is also especially pertinent in the case of the Madres. For the Madres, maternal imagery was initially a means to protest against authoritarianism. Yet, as noted in chapter 9, this imagery evolved. In the beginning, the Madres were mobilized by a traditional view of the sexual division of labor. Several years later, however, the Madres said that they "awakened." They were comfortable meeting with newspaper reporters and with the president of Mexico. This process of identity transformation has also kept them involved in the movement. As Mirna Benítez notes in chapter 9, each of the Madres came to the movement as a result of different interests, yet *now* they understand the connection between the personal and the political—from their homes, where they are mothers, to the nuclear industry and the political system beyond.

Although many of the NSM concepts appear to apply to the Mexican antinuclear movement, one New Social Movement concept clearly does not: the Mexican antinuclear groups cannot claim to be pursuing purely postmaterial goals. Alberto Melucci views NSMs as movements concerned with identity issues. Most of the participants in these movements are either middle-class individuals or students. Hence, their motivation for joining these movements is postmaterial: that is, unlike members of the labor movement, these political actors have mobilized to achieve noneconomic goals. Further, also unlike the labor movement, NSMs do not necessarily "require engagement or confrontation with the institutions of the State."[4]

This aspect of NSM theory does not apply in the case of the Mexican antinuclear movement. The cattlemen from Palma Sola joined the movement primarily to defend their source of income; they were afraid that no one would buy their meat, milk, and other agricultural products out of fear of radioactive contamination. The fishermen joined because their catches were plummeting and they had no other source of income. Even in the case of the upper-middle-class and upper-class Madres Veracruzanas, some economic goals were important. At least one of the Madres from the port of Veracruz was defending her expensive beachfront home, located only a few miles from the Laguna Verde plant.

Further, the Mexican antinuclear movement, though it sought to maintain its autonomy from the political system and parties, nevertheless had to engage with the state in order to close Laguna Verde. The Madres Veracruzanas, for example, vehemently protected their autonomy while simultaneously meeting with government officials, including the president. They believed that, given Mexico's centralized political system, it was simply impossible to achieve their goal without engaging the state in some way.

The NSM approach, then, does not account well for certain aspects of the Mexican antinuclear power movement. For example, presently very little antinuclear activity is occurring in the Palma Sola region. An NSM approach might look for latency and low-level participation in the submerged networks. However, in this case it appears that the repression and co-optation tactics used by the Mexican state have been so overwhelming that the villagers have become disillusioned and discouraged to the point of inactivity. Indeed, one former activist believes that the town never recovered from its "collective depression."[5] Therefore, in the case of Palma Sola, the NSM approach is not entirely useful in understanding the villagers' experiences and the change in the level of their involvement in the movement.

The New Social Movements approach is also inadequate for analyzing the related phenomenon of censorship and its effects on a movement. With its focus on the construction of identity and meaning, the approach does not account sufficiently for external forces. For example, the cattlemen and fishermen's revelations about the presence of carcinogens in the Gulf waters and about the problems in the way that the plant was run were not carried by television news programs. Further, the constant protest marches and blockades of traffic throughout 1988 and 1989 were never covered by television news. The NSM literature—heavily influenced by studies of movements in Europe and the United States, where the television news is not censored—has not provided a framework for the study of censorship.

In order to remedy these limitations of the NSM framework, Arturo Escobar and Sonia Alvarez suggest that students of Latin American social movements combine the resource mobilization model with the New Social Movements perspective (see chapter 1).[6] I would argue, however, that the NSM perspective would work better in combination with the *political process model*.

Political Process and State-Movement Relations

The political process model is valuable because, unlike the New Social Movements and resource mobilization models, it looks beyond the social movement itself to the political context of mobilization. The best-known version of political process has been put forth by Doug McAdam in his book *Political Process and the Development of Black Insurgency, 1930–1970*.[7] Like resource mobilization, political process sees social movements as inherently political, rather than psychological, phenomena. But unlike resource mobilization, the political process approach analyzes the ideology guiding a movement instead of treating the movement as merely a product of its resources.[8]

The political process approach argues that environmental circumstances are important and can change over time. McAdam believes that the larger political context can prevent mass mobilization, but at certain times "political opportunities" arise that permit enough mobilization to allow for the creation of a social movement.[9] In other words, the political system can be more amenable to a social movement's demands at certain times than at others. According to the political process model, the political context has to be analyzed; that is, it is not a neutral background for a social movement.

In addition, the political process approach analyzes the nature of the movement itself. McAdam believes that a propitious context is not enough; the movement must also have access to resources and must maintain the movement's "organizational strength," which has various aspects.[10] It lies partly in the membership of the organizations that participate in the movement; for McAdam "mobilization does not occur through recruitment of large numbers of isolated and solitary individuals. It occurs as a result of recruiting blocs of people who are already highly organized and participants."[11] A movement's strength also depends on the ability of organizations to provide members with incentives for participation. Thus, providing members with "interpersonal rewards"[12] can solve the "free rider" problem that is so important to analysis in the resource mobilization paradigm. A movement's strength, finally, depends on an efficient communication network and on good leaders who can take advantage of political opportunities.[13]

Concepts such as "political opportunities" and the "social control response" of the state are useful in understanding state-movement relations in the Mexican case. The concept of political opportunities refers to the fact that state structures are not static; they can be more or less amenable to a movement's demands at different points in time. It also refers to the degree of openness of a particular political system. In other words, the nature of political structures conditions how grievances are manifested and how movements choose to mobilize.

Christian Joppke, using a "contextualized political process perspective,"[14] sees both state and movement as constantly changing. Neither is inert, and their interaction affects each actor. In his analysis of the West German antinuclear movement, Joppke argues that the movement's strategies can be explained by examining political opportunities. The German state has historically been much more centralized than that of the United States. Because Germany was a late industrializer, the state had to play a direct role in the promotion of economic development. Thus, it was much more centralized and interventionist than the U.S. government. In addition, the German state was solidly behind the use of nuclear energy. As a result, the German antinuclear movement was different from its U.S. counterpart. Because political power is more widely dispersed in the United States, the U.S. antinuclear movement was more fragmented and regional. In Germany, on the other hand, the state turned nuclear installations into "paramilitary fortresses"[15] and was very aggressive in its treatment of the movement, which was then almost forced to become antistatist. Many antinuclear groups in the United States, on the other hand, could use conventional political channels—along with NSM tactics—in order to achieve their goals through the use of existing political institutions.

The Mexican case is somewhat similar to the German one.[16] Mexico's political system is centralized and hierarchical: the executive branch dominates the judicial and legislative branches of government. Moreover, one party, the PRI, has been in power since its creation in 1929. The political system is thus corporatist and authoritarian. In addition, because the Mexican state, like the German state, has been very much in favor of nuclear energy, the Mexican antinuclear movement—like its German counterpart—has had to emphasize confrontational tactics in its strategies. In the Mexican case, the most visible antinuclear activities were protests and highway blockades. These tactics got the state's attention: after many of these protests, government officials offered to meet with the antinuclear

groups. However, as in the German case, the Laguna Verde nuclear installations and surrounding area also became heavily militarized. The village
of Palma Sola was essentially in a state of siege; the area was patrolled by
the military, residents had to pass through a military checkpoint when
leaving the village, and warships were visible just off the coast. Although
the Mexican movement did attempt to use conventional political channels at times—for example, filing injunctions in the court system—this
strategy was completely unsuccessful. Judge after judge ruled against the
movement. In sum, political opportunities for the Mexican antinuclear
movement were more limited than those available to the U.S. movement.
The Mexican state favored nuclear energy, and the PRI was not interested
in allowing antinuclear arguments within its platform, which meant that
organizations within the movement had to work outside normal political
channels in order to get their message across to Mexican society.

The political process model views the state not only as the creator of
political conditions affecting a movement but also as a political actor that
actually interacts with the social movement. Certainly the state was very
active in the Mexican case. As described in chapters 2 and 3, the Mexican
state responded to the antinuclear movement in a variety of ways—by
censoring television news, by engaging in tactics of repression against individuals and groups, and by attempting to co-opt individual movement
leaders as well as the residents of the village of Palma Sola. These state
actions had a profound impact on the movement. In 1988, when soldiers
aimed their weapons at women and children who were attempting to march
out of Palma Sola toward the plant, the state's show of strength so increased the villagers' fears that they completely dropped out of the movement. The thought that the military could actually threaten to harm unarmed and nonviolent women and children was too much for the villagers
to bear.

The state's co-optation tactics also deeply affected the rest of the movement. The cattlemen of Palma Sola, one of the most active antinuclear
groups in Mexico, were devastated when one of their leaders was recruited
by the PRI in 1989. They became so disillusioned by this turn of events
that they too dropped out of the movement shortly thereafter. The Madres
Veracruzanas expelled their leader in 1988 because they suspected that
she was growing too close to the government when her husband accepted
a government position in Coatepec, Veracruz. These events came to loom

large in the movement. Subsequently, movement participants often looked at fellow members with suspicion, wondering which participant would be co-opted next.

Thus, the political process model in this instance complements the New Social Movements approach very well. The NSM literature focuses on how New Social Movements tend to build solidarity. It views these groups as nonhierarchical and fluid in structure, seeking to achieve nonmaterialist goals, which often involve changes in identity. Certainly many of the antinuclear groups in Mexico can be at least partially described in this way. However, because many NSM groups profess a desire to maintain their distance from the state, the NSM approach cannot be used to analyze state-movement relations, which means that the Mexican case needs to be analyzed from the perspective of a paradigm that takes these relations into consideration. While the Mexican antinuclear groups attempted to promote solidarity within their nonhierarchical, grassroots organizations, the Mexican state was busily responding in ways that often drove wedges between participants. Thus, the political process approach is on target when it emphasizes the idea that state and movement are not static and that their interactions affect each actor.

It is important to recognize, however, that the political process approach, like the New Social Movements approach, was created to analyze movements in advanced, industrialized, democratic societies. Certain aspects of this approach, then, must be modified to fit authoritarian contexts, such as Mexico. Specifically, some scholars studying social movements in authoritarian political environments have analyzed the phenomenon of the elites' tolerance for independent movements, an important component of the concept of "political opportunities."[17] In the case of the independent teachers' movement in Mexico, a national-level elite conflict between union and government officials created the political space necessary for the movement to emerge. Subsequently, a few regional-level teachers' movements sustained their activities through "an elaborate network of grass-roots organizations."[18] They also successfully maintained their democratization efforts by avoiding affiliation with particular political parties.

In the case of the Mexican antinuclear movement, the government's tolerance was multidimensional. Although the Madres Veracruzanas' organizing activities were tolerated, the Palma Sola villagers' activities were not. At times, members of CONCLAVE were also beaten and harassed. A similar phenomenon occurred in Brazil in the 1980s, when members of women's organizations did not experience the level of repression aimed at other organizations.[19] The concept of the elite's tolerance appears to have a gendered component in many Latin American societies: women, especially mothers, are regarded as "apolitical" and thus escape the brunt of the repression aimed at many members of social movements.[20]

The Mexican Antinuclear Movement and the Question of Democracy

The competitive 1988 presidential election, combined with a series of electoral reform laws dating back to the 1970s, led many analysts to assume that the Mexican political system was in the process of becoming more democratic. President Luis Echeverría (1970–76) engaged in a "democratic opening" that allowed independent unions to organize and strike and that also lessened censorship of the press.[21] Later, President José López Portillo (1976–82) oversaw a political reform process that allowed for the formation of a political opposition within the framework of the electoral process. The 1977 law allowed for 25 percent of the seats in the Chamber of Deputies to be reserved for minor parties. Nevertheless, it still "guarantee[d] the continued numerical superiority of the official party in all governing bodies, local, state, and federal."[22]

The presidential election of 1988 also led many analysts to assume that Mexico was becoming more democratic. In that election Carlos Salinas of the PRI ran against Cuauhtémoc Cárdenas of the left-leaning Frente Democrático (later the PRD) and against Manuel Clouthier of the right-wing PAN. Compared to previous elections (in some of which the PRI candidate essentially ran unopposed), the 1988 election was extremely competitive. In addition, "modern" political practices such as polling became important for the first time; polls showed that Cárdenas and Clouthier had much support throughout the country. The election was held on July 6, 1988, but election results were not available for several days. Soon after the election, Miguel Bartlett, the minister of the Secretaría de Gobernación, announced that the computers counting the votes had broken down,[23] which, of course, raised suspicions that the PRI was using new technology to engage in its age-old practice of electoral fraud. After

July 8, both Salinas and Cárdenas claimed to be the winner. Finally, the Election Commission announced that Salinas had won with 50.7 percent of the vote, compared with Clouthier's 16.8 percent and Cárdenas's 31.1 percent. These results were far different from those of previous presidential elections, in which the PRI candidate generally won by an overwhelming majority.[24]

Does this mean that Mexico is on the road to democracy? Many students of Mexican politics are very skeptical. Judith Adler Hellman, for example, notes that

> The PRI is clearly disposed to negotiate with the opposition on the rules of the political game but not on the results. The appropriation by the ruling party of what was most likely an opposition victory makes clear that the official party is not prepared to cede presidential power under any circumstances. And, in the elections held since 1988, evidence of electoral fraud continues to be incontrovertible. In short, since coming to power, the Salinas regime has employed harassment, intimidation, manipulation, and outright fraud to guarantee that the PRI would regain lost ground.[25]

This pattern has not changed significantly under President Ernesto Zedillo. Indeed, current scandals involving political assassinations, politicians' cooperation with drug traffickers, and electoral fraud have tainted Mexico's image internationally.

Did the Mexican antinuclear movement, then, push the Mexican political system in a more democratic direction? Given the fact that the Mexican government has not closed down the Laguna Verde plant, the answer has to be no. When the movement first emerged, it was clear that the antinuclear groups did not enjoy the ample political opportunities available in Western industrialized democracies. Further, as discussed previously, the Mexican government did not simply turn a deaf ear to the movement's demands: it combined co-optation tactics with instances of repression in order to weaken and silence the movement. Although the government's response varied depending on the class and gender of the participants, actions such as beatings of CONCLAVE members and threatening telephone calls were designed to intimidate the membership of the movement as a whole. Similarly, the co-optation of the leader of the Cattlemen's Association led to the cattlemen's withdrawal from the movement. The state also censored television news coverage. Finally, the anti-nuclear groups often accused the Mexican government of being authoritarian: they repeatedly demanded a referendum on the nuclear energy issue, but the Mexican government declined the request. The evidence therefore does not indicate that the antinuclear movement pushed the Mexican government in a more democratic direction.

The antinuclear groups found it difficult to overcome these impediments to mobilization, but it would be wrong to suggest that their activities had no effect on society or politics. The movement continued when various groups served as watchdogs: they revealed that the evacuation plan was woefully inadequate and that minor accidents had occurred at Laguna Verde. Initially, the government denied that these problems existed but was forced to acknowledge them after antinuclear activists publicly announced that the incidents had occurred. The antinuclear groups won these small victories despite all the obstacles presented by the government.

Further, I would argue that though the movement failed to close Laguna Verde, some of the organizations—especially the Madres Veracruzanas—helped to strengthen Mexican civil society. María Lorena Cook refers to the Mexican democratic teachers' movement's creation of "potential 'citizens' where before there had been clients or masses."[26] Similarly, as described in chapter 9, the Madres Veracruzanas argue that they experienced a profound transformation as a result of their participation in the movement. The group struggled to conduct its business democratically and chose to have no formal leaders. They also learned a great lesson in the ways of Mexico's authoritarian politics. All of the Madres now recognize the limitations of the current regime and support greater democratization—especially in the area of high technology. Thus, even "failed movements" can contribute to a democratic undercurrent in civil society, which can become a torrent in future circumstances.

Finally, the case of the Madres Veracruzanas serves to illuminate the complexity of the feminist debate concerning mothers' movements. Some feminists argue that mothers' movements are misguided because they serve to reinforce the sexual division of labor in society. Others romanticize mothers' movements. But in the Mexican case it is clear that maternal imagery is not necessarily static. The Madres Veracruzanas have changed as a result of their participation in the group and in the larger environmental struggle. Thus, although mothers' movements may appear to be purely traditional in that they reinforce the sexual division of labor, they have the potential to become multidimensional and to change their participants. In addition, the case of the Madres Veracruzanas indicates that mothers' movements should not be romanticized: the Madres have erected class barriers that have prevented the recruitment of women who do not belong to the middle class. Despite this problem and despite the fact that the Madres Veracruzanas have not achieved their goal of closing the Laguna Verde plant after years of participating in the movement, they continue to keep struggling. They are the only remaining active antinuclear group in Mexico.

Appendix A

Chronology of the Mexican Antinuclear Movement

1965   Administration of President Gustavo Díaz Ordaz decides to embrace nuclear energy.

1972   Mexican government, after a bidding process, signs a contract with General Electric to build a nuclear power plant.

1981   Purépecha grassroots movement blocks the construction of an experimental nuclear reactor in the state of Michoacán.

1986   Numerous groups, including the Madres Veracruzanas, are founded to prevent the Mexican government from operating the Laguna Verde plant.

April 26:   accident at Chernobyl.

1988   June 5:   PAN holds a referendum in Veracruz on the nuclear issue.

June 19–22:   Longest of several highway blockades erected by antinuclear groups to protest Laguna Verde; military forcefully removes demonstrators at Palma Sola.

July 6:   In the presidential election, Carlos Salinas de Gortari is declared winner over Cuauhtémoc Cárdenas.

July 26:   Government announces Laguna Verde will not operate "for the moment."

October 13:   Senate approves operation of Laguna Verde.

October 15:   President Miguel de la Madrid approves the operation of Laguna Verde before the lower half of Chamber of Deputies begins its debate.

October 18:   Lower half of Chamber of Deputies approves Laguna Verde.

October 21:   Women and children threatened by military in Palma Sola.

November 16:   Villagers of Palma Sola hear explosion; government first says it is caused by an emergency drill but later admits the plant experienced a minor accident.

	December: Madres Veracruzanas expel their leader, Rebeka Dyer.
1989	February 18: CONCLAVE is founded and brings together many smaller groups.
	August: CFE accuses Miguel Angel Valdovinos Terán of passing confidential information to antinuclear activists and relieves him of his duties as an engineer at Laguna Verde.
	August: Efrén López Meza, the cattlemen's leader, defects and accepts political positions with the PRI.
	November: Madres Veracruzanas meet with President Carlos Salinas de Gortari, who promises them an independent audit of Laguna Verde.
1990	March: Confidential fax made public; the Laguna Verde audit would be rigged.
	August: Outside evaluators declare that Laguna Verde has no problems and can operate.
1991	July: Madres meet with Miguel Alemán Velasco, candidate for the Senate.
1993	January: Article by journalist Guillermo Zamora in the weekly *Proceso* maintains that Laguna Verde has serious problems and that the evacuation plan is deeply flawed.
1994	July: Mirna Benítez of the Madres Veracruzanas runs unsuccessfully for the Senate.
1996	April: Madres join Greenpeace at a protest at Laguna Verde to commemorate the tenth anniversary of the Chernobyl nuclear accident.

Appendix B

The Mexican Antinuclear Organizations

Organizations	Leaders	Period of Activity
The Most Active Antinuclear Groups		
Madres Veracruzanas	Rebeka Dyer	1986–present
	Mirna Benítez	
Grupo de los Cien	Homero Aridjis	1986–present
	Feliciano Béjar	
Cattlemen's Association	Efrén López Meza	1987–1992
	Jesús Rodal Morales	
CONCLAVE	Pedro Lizárraga	1989–present
	Juan Marín	
Veracruz Fishermen	Eduardo Gómez Téllez	1988–1991
Pacto de Grupos Ecologistas	Víctor Meza	1986–1989 (joined CONCLAVE)
	Marco Martínez Negrete	
Other Antinuclear Groups		
Grupo Arcoíris	Mariano López	1987–1991
Frente Regional Antinuclear		1987–1989
Asociación Ecológica del Istmo		1987–1989
Grupo Quetzalcoatl	José Carrasco Flores	1987–1989
Grupo Antinuclear de Coatepec	Fernando Jácome	1987–1989
Movimiento Nacional Ecologista		1986–1989
Partido Verde Mexicano	Jorge González Torres	1986–1989

During the late 1980s many antinuclear groups emerged to protest the Laguna Verde nuclear power plant. The year 1989, however, marks the demise of many groups, for two reasons: most of them simply joined CONCLAVE and disappeared as independent entities; others (e.g., the Grupo Antinuclear de Coatepec) simply dissolved.

Notes

Chapter 1. Introduction

1. I personally witnessed this incident.

2. Fernando Calderón, Alejandro Piscitelli, and José Luis Reyna, "Social Movements: Actors, Theories, and Expectations," in Arturo Escobar and Sonia Alvarez, eds., *The Making of Social Movements in Latin America: Identity, Strategy, and Democracy* (Boulder, Colo.: Westview, 1992).

3. Arturo Escobar, "Imagining a Post-Development Era? Critical Thought, Development and Social Movements," paper prepared for the international seminar CENDES, October 1–6, 1990, Universidad Central de Venezuela, Caracas. For an overview of the New Social Movements literature see, for example, Bert Klandermans and Sidney Tarrow, "Mobilizing into Social Movements: Synthesizing European and American Approaches," in Hanspeter Kriesi, Sidney Tarrow, and Bert Klandermans, eds., *International Social Movements Research* (London: JAI, 1988); and David Slater, ed., *New Social Movements and the State in Latin America* (Amsterdam: CEDLA, 1988). Empirical studies on Latin America include María del Pilar García, *Estado, ambiente, y sociedad civil* (Caracas: USB/ CENDES, 1991). Empirical studies on the United States and France include Jerome Price, *The Antinuclear Movement* (Boston: Twayne, 1982), and Alain Tourraine, *Antinuclear Protest: The Opposition to Nuclear Energy in France* (New York: Cambridge Univ. Press, 1983).

4. Alan Scott, *Ideology and the New Social Movements* (London: Unwyn Hyman, 1990), 17.

5. Ibid.

6. Alberto Melucci, *Nomads of the Present: Social Movements and Individual Needs in Contemporary Society,* ed. John Kean and Paul Mier (Philadelphia: Temple Univ. Press, 1989), 41.

7. Melucci, *Nomads of the Present,* 60.

8. Ibid.

9. Ibid.

10. Ibid.

11. Calderón, Piscitelli, and Reyna, "Social Movements," 27.

12. Ibid.

13. Melucci, *Nomads of the Present,* 4.

14. Ibid., 35.

15. Klandermans and Tarrow, "Mobilizing into Social Movements," quoted in Alvarez and Escobar, eds., *The Making of Social Movements in Latin America,* 317–18.

16. Alvarez and Escobar, eds., *The Making of Social Movements in Latin America,* 319–20.

17. For an overview of the resource mobilization paradigm, see J. D. McCarthy and M. N. Zald, eds., *Social Movements in an Organizational Society* (New Brunswick, N.J.: Transaction Books, 1977).

18. McCarthy and Zald, *Social Movements,* 122, quoted in Doug McAdam, *Political Process and the Development of Black Insurgency, 1930–1970* (Chicago: Univ. of Chicago Press, 1982), 32.

19. Because the pluralist view of the political system is that power is dispersed and access points are plentiful, those actors who resort to activity outside the system must be succumbing to psychological pathologies.

20. McAdam, *Political Process,* 22–23.

21. The military sector of the party was eliminated in the 1940s.

22. Ruth Berins Collier and David Collier, "Inducements versus Constraints: Disaggregating Corporatism," *American Political Science Review* 73 (1979), 968. For example, peasants receiving land under the agrarian reform program after the revolution were obliged to join the Confederación Nacional Campesina, the peasant sector of the party.

23. Alfred Stepan, *The State and Society: Peru in Comparative Perspective* (Princeton: Princeton Univ. Press, 1978); John Sloan, "The Mexican Variant of Corporatism," *Inter-American Economic Affairs* 38 (spring 1985), 3–18.

24. Sloan, "Mexican Variant," 17.

25. Peter H. Smith, "Does Mexico Have a Power Elite?" in José Luis Reyna and Richard S. Weinart, eds., *Authoritarianism in Mexico* (Philadelphia: Institute for the Study of Human Issues, 1977), 146, quoted in Sloan, "Mexican Variant," 17.

26. Diane Davis, "Failed Democratic Reform in Mexico: From Social Movements to the State and Back Again," *Journal of Latin American Studies* 26 (1994), 375–408.

27. Joe Foweraker, "Introduction," in Joe Foweraker and Ann Craig, eds., *Popular Movements and Political Change in Mexico* (Boulder, Colo.: Lynne Rienner, 1993), 7.

28. Ibid., 8–9.

29. Vivienne Bennett, "The Evolution of Urban Popular Movements in Mexico between 1968 and 1988," in Alvarez and Escobar, eds., *The Making of Social Movements in Latin America,* 240.

30. Bennett, "The Evolution of Urban Popular Movements," 247.

31. Ibid., 254.

32. Diane Davis, "Failed Democratic Reform in Mexico," 376.

33. Foweraker, "Introduction," 11.

34. Judith Adler Hellman, "Mexican Popular Movements, Clientelism, and the Process of Democratization," *Latin American Perspectives* 81, no. 21.2 (1994), 128.

35. Paul Haber, "Identity and Political Process: Recent Trends in the Study of Latin American Social Movements," *Latin American Research Review* 31 (1996), 181.

36. By comparison, Mexico's "popular" movement demands typically include material issues—land, jobs, and housing.

37. Jonathan Fox, "The Difficult Transition from Clientelism to Citizenship, Lessons from Mexico," *World Politics* 46, no. 2 (1994), 151–52.

38. Ibid., 152.

Chapter 2. The Origins of the Mexican Antinuclear Movement

1. Hugo García Michel, *Más allá de Laguna Verde* (Mexico City: Editorial Posada, 1988), 102.

2. Carmen Buerba, "De Pátzcuaro a Laguna Verde: La experiencia antinuclear en Santa Fé de la Laguna," in José Arias Chávez and Luis Barquera, eds., *Laguna Verde Nuclear? No Gracias* (Mexico City: Claves Latinoamericanas, 1988), 245.

3. Ibid., 248.

4. Victor Paya Porres, *Laguna Verde: La violencia de la modernización* (Mexico City: Miguel Angel Porrúa, 1994), 55.

5. For a history of Mexican environmentalism, see Lane Simonian, *Defending the Land of the Jaguar: A History of Conservation in Mexico* (Austin: University of Texas Press, 1995).

6. Stephen P. Mumme, "System Maintenance and Environmental Reform in Mexico: Salinas's Preemptive Strategy," *Latin American Perspectives* 72, no. 19.1 (1992), 124.

7. Ibid.

8. Ibid., 127.

9. Ibid.

10. Guillermo Zamora, "Todas las clases sociales de Veracruz contra Laguna

Verde," *Proceso,* May 1987, 18–19.

11. See David De Leon, *Everything Is Changing: Contemporary U.S. Movements in Historical Perspective* (New York: Praeger, 1988), esp. chapters 1 and 2.

12. Antonio Bretón, former treasurer of the local cattlemen's association, explained to me that the evacuation procedure's map is not accurate. Entire villages are missing from the map, and the major evacuation routes are, in reality, barely passable dirt roads. Interview with Antonio Bretón, July 21, 1988, Palma Sola, Veracruz. For a detailed critique of the evacuation plan, see Alejandro Nadal and Octavio Miramontes, "Análisis crítico: El plan de emergencia radiológica externo de Laguna Verde," *La Jornada,* October 16, 1988, 15–18.

13. Interview with Feliciano Béjar, February 22, 1990, Mexico City.

14. Ibid.

15. Her child's godmother.

Chapter 3. The Mexican Antinuclear Power Movement, 1987–1988

1. "En Veracruz se reítera el respeto a toda ideología: FGB," *Diario del Istmo,* February 20, 1988, 1.

2. Ibid.

3. "Advierten los opositores a la nucleoeléctrica," *Diario de Xalapa,* February 25, 1988, 1.

4. Ibid.

5. Ibid.

6. G. García Rivera, "Levantan el bloqueo de carreteras," *Diario del Istmo,* February 26, 1988, 1.

7. Ibid. All translations are mine.

8. Roberto Sosa, "Bloquearon una carretera en Minatitlán," *Excélsior,* February 29, 1988, Sección en los Estados, 1.

9. "En protesta contra Laguna Verde," *Diario de Xalapa,* March 1, 1988, 2.

10. Interview with the Madres Veracruzanas, June 25, 1988, Xalapa, Veracruz.

11. Ibid.

12. Ibid.

13. Interview, March 6, 1988.

14. Ibid.

15. Miguel Angel Rivera, "Gutiérrez Barrios: No es irreversible la puesta en marcha de Laguna Verde," *La Jornada,* March 7, 1988, 8.

16. Interview with Antonio Bretón, July 21, 1988, Palma Sola, Veracruz.

17. Ibid.

18. Laura Robles, "Realizarán dos marchas de protesta en contra de la planta

nuclear Laguna Verde," *Gráfico de Xalapa,* June 9, 1988, 1.

19. Elvira Marcelo Esquivel, "Pide el pacto ecologista la cancelación de Laguna Verde," *El Día,* June 17, 1988, 7.

20. Ibid.

21. Joaquín Rosas Garcés, "Opinión," *El Dictamen,* June 20, 1988, 1.

22. Marco Antonio Aguirre, "No hay paso en Palma Sola: Agua y comida para muchos días," *El Dictamen,* June 20, 1988, 1.

23. Quoted in ibid.

24. Rosalinda Sáenz y Zárate, "Ya cumplieron 46 horas: Sigue bloqueada la costera en protesta contra el funcionamiento de L.V.," " *Gráfico de Xalapa,* June 21, 1988, 1.

25. Ibid.

26. Gabriel Arellanos López, "'Ilegal' califica F.G.B. la actitud de los ecologistas," *Grafico de Xalapa,* June 21, 1988, 1.

27. Ibid.

28. "Por L.V., más de un millón de Mexicanos en peligro," *Diario del Istmo,* June 21, 1988, 1 and 12A.

29. Ibid.

30. Paid advertisement, *Diario de Xalapa,* June 22, 1988, 10. According to Rebeka Dyer, some members of the Madres Veracruzanas participated in blockades, though the group never organized these activities. The Madres also gave blockade participants any inside information they managed to gather from governmental sources. Interview with Rebeka Dyer, June 25, 1995, Xalapa, Veracruz.

31. Ibid.

32. Ana Luisa Murrieta and Daniel Ruíz, "Se agrava el caos por el bloqueo carretero," *Diario de Xalapa,* June 22, 1988, 1.

33. Ibid.

34. Ibid.

35. Alfonso Valencia Ríos, "Tensa la situación de Palma Sola: Los traileros decidiosos a librar el paso," *El Dictamen,* June 22, 1988, 1.

36. Ibid.

37. Orlando García and Alma Elena Gutiérrez, "Pérdidas diarias de mil millones por un bloqueo carretero," *Excélsior,* June 22, 1988, Sección en los Estados, 1.

38. Ibid.

39. Interview with Marta Lilia Aguilar, October 21, 1988, Rancho Brazo Fuerte, Palma Sola, Veracruz.

40. Héctor Ramos, "Posible acción militar para desalojar a ecologistas," *Diario de Xalapa,* June 23, 1988, 1.

41. Ibid. Gobernación is roughly equivalent to the Department of the Interior

in the United States.

42. Ibid.

43. Orlando García, "Desbloquearon la carretera opositores a Laguna Verde," *Excélsior,* June 24, 1988, Sección en los Estados, 1.

44. Ibid.

45. Ibid.

46. Rosa Rojas, "Violencia en Palma Sola, acusan," *La Jornada,* June 25, 1988, 9.

47. Ibid.

48. Ibid.

49. *Ejidatarios* are peasant farmers who received lands through the agrarian reform program after the Mexican Revolution of 1910–20.

50. Martha Meza, "Recibirá M.M.H. a los antinucleares," *Sol Veracruzano,* June 28, 1988, 1.

51. Francisco Rivera Palacios, "Rechazo a L.V. y republicación de el Sardinero," *El Dictamen,* June 25, 1988, 5.

52. Ibid.

53. Interview with the Madres Veracruzanas, June 25, 1988, Xalapa, Veracruz.

54. Ibid.

55. Luz María Rivera, "No hay posibilidades de un referéndum sobre Laguna Verde: Lizárraga," *Sol Veracruzano,* May 17, 1988, 1.

56. Ibid.

57. Ibid.

58. María Elena Hernández, "Bloquerán las carreteras del Estado de Veracruz dos horas," *El Dictamen,* May 26, 1988, 1.

59. Ibid.

60. Alma Elena Gutiérrez, "Moños rojos contra la nucleoeléctrica," *Excélsior,* June 1, 1988, Sección en los Estados, 1.

61. Evangelina Hernández, "Editorial," *La Jornada,* June 3, 1988, 1.

62. Antonia Castillo Santos, "Hoy, el referéndum sobre el funcionamiento de Laguna Verde," *El Dictamen,* June 5, 1988, 1.

63. Ibid.

64. Interview with the Madres Veracruzanas, June 25, 1988, Xalapa, Veracruz.

65. Ibid.

66. Rosa Elvira Vargas, "Editorial," *El Financiero,* June 13, 1988, 1.

67. Ibid. The population of Xalapa is approximately 500,000.

68. The PAN historically has had more support in northern states such as Chihuahua. Silverio Quevedo Elox, "Fracasó el referéndum del PAN sobre L.V. en Veracruz," *Diario del Istmo,* June 6, 1988, 1.

69. Carlos Jesús García, "El referéndum, la peor farsa del PAN: Jorge Moreno

Salinas," *El Dictamen,* June 7, 1988, 1.

70. Marco Antonio Aguirre, "El referéndum, madurez cívica, rechazo popular a Laguna Verde," *El Dictamen,* June 8, 1988, 1.

71. The Partido de la Revolución Democrática was founded by Cuauhtémoc Cárdenas and other defectors from the PRI. This group, known as the Democratic Current, left the PRI because it believed that the party was not pursuing democracy vigorously enough.

72. Interview with the Madres Veracruzanas, June 25, 1988, Xalapa, Veracruz. The Madres refused to support Cárdenas openly despite his antinuclear stance.

73. Ibid.

74. Carlos de Jesús Rodríguez, "La defensa del voto hasta las últimas consecuencias: Cuauhtémoc," *El Dictamen,* July 24, 1988, 1.

75. "El Reactor de Laguna Verde no se cargará," *Excélsior,* July 27, 1988, 5A.

76. Eibenschutz, a Mexican engineer, is known as "the father of Laguna Verde," and during this period he was a member of the Federal Electricity Commission, which oversees Laguna Verde's operations. "Eibenshutz de la CFE se opone a retardar la operación de Laguna Verde," *El Día,* July 28, 1988, 1.

77. Interview with Fernando Jácome, July 28, 1988, Xalapa, Veracruz.

78. Ibid.

79. Open letter, Madres Veracruzanas, August 4, 1988.

80. Carlos Martínez Rentería, "Quiero el cambio de México, dice Béjar," *Diario del Istmo,* August 1, 1988, 12.

81. Ibid.

82. Ibid.

83. "Tememos que la carga de Laguna Verde sea después de los comicios," *Diario de Xalapa,* August 1, 1988, 1.

84. Ibid.

85. Elvira Marcelo Esquivel, "Consecuencias desastrosas al medio ambiente si no se pone en operación Laguna Verde: CFE," *El Día,* August 6, 1988, 8.

86. Ibid.

87. Ibid.

88. Fany Yépez, "Necesario el funcionamiento de Laguna Verde: Alcudia García," *Sol Veracruzano,* August 7, 1988, 1.

89. Ibid.

90. Ibid.

91. Rosa Contreras, "Por no funcionar L.V., se pierden al día 180 millones de dólares: Carlos Smith," *Diario de Xalapa,* August 7, 1988, 1.

92. Ibid.

93. Ibid.

94. Personal communication with the Madres Veracruzanas, August 8, 1988,

Austin, Texas.

95. Ibid.

96. Ibid.

97. Ibid.

98. Ibid.

99. Daniel Ruíz, *Diario de Xalapa,* August 9, 1988, 1.

100. Ibid.

101. Ibid.

102. Ibid.

103. Manuel Nóguez Viguera, "Hay fallas graves en Laguna Verde; La amenaza es constante, dicen ecologistas," *Excélsior,* August 20, 1988, 5A.

104. Ibid.

105. Ibid.

106. Gustavo Esteva, "Editorial," *El Día,* August 21, 1988, 1.

107. Ibid.

108. Ibid.

109. Ibid.

110. Bertha López Aguayo, "Preocupante, la indefinición sobre Laguna Verde," *Diario de Xalapa,* September 3, 1988, 1.

111. Ibid.

112. Ibid.

113. Mario Jareda, "Factible la conversión de la planta de Laguna Verde," *Diario de Xalapa,* September 5, 1988, 1.

114. Violeta Pacheco García, "Lombardo declara: Laguna Verde será puesta en operación," *Diario de Xalapa,* September 21, 1988, 1.

115. Ibid.

116. "Lombardo desorienta a la opinión pública," *Diario de Xalapa,* September 22, 1988, 1.

117. Quoted in ibid.

118. Emilio Lomas, "Laguna Verde podría operar en este sexenio," *La Jornada,* September 22, 1988, 23.

119. Aníbal Ramírez, "Invita la CFE a ecologistas para que comprueben la seguridad de Laguna Verde," *El Día,* September 23, 1988, 10.

120. Ibid.

121. Homero Aridjis, "Contra la censura sobre Laguna Verde," *La Jornada,* September 27, 1988, 13.

122. Homero Aridjis in ibid.

123. Homero Aridjis in ibid.

124. Sonia García, "Invitan las Madres Veracruzanas a manifestarse contra Laguna Verde," *Sol Veracruzano,* September 29, 1988, 1.

125. M. A. Aguirre, "Contra Laguna Verde se pronuncia Carmen Cover," *El*

Dictamen, September 29, 1988, 5A.

126. Quoted in ibid.

127. Alma Elena Gutiérrez, "Laguna Verde sin fecha para operar: M. L. Fuentes," *Excélsior,* September 29, 1988, Sección en los Estados, 1.

128. Quoted in ibid.

129. Ibid.

130. Quoted in Victor Ballinas, "Que se realice un referéndum en Veracruz," *La Jornada,* September 29, 1988, 21.

131. Ibid.

132. Juan Víctor Artega, "El Pueblo no se opone a que opere Laguna Verde," *El Día,* October 1, 1988, 8.

133. Ibid.

134. Quoted in ibid.

135. Carlos Medina, "También a fugas radiactivas: R. González," *Excélsior,* October 1, 1988, 4A.

136. Quoted in ibid.

137. Quoted in ibid.

138. Emilio Lomas, "El permiso está listo, informó la CNSNS y S," *La Jornada,* October 6, 1988, 7.

139. "Avala el SUTERM la operación inmediata de Laguna Verde," *La Jornada,* October 7, 1988, 7.

140. Avelino Hernández Vélez, "Editorial," *El Financiero,* October 7, 1988, 1.

141. Ibid.

142. "Laguna Verde no debe entrar en operación; Es un atentado contra el pueblo: Los Cien," *Diario de Xalapa,* October 7, 1988, 1.

143. Ibid.

144. Ibid.

145. Ibid.

146. Ibid.

147. "Acuerdo del gobernador y líderes," *El Dictamen,* October 7, 1988, 1.

148. Quoted in ibid.

149. Quoted in ibid.

150. Quoted in ibid.

151. Madres Veracruzanas, paid political advertisement, *Diario de Xalapa,* October 1, 1988, 14.

152. Arturo Zárate Vite, *El Universal,* October 2, 1988.

153. Announcement, Movimiento Antinuclear y Ecologista, October 3, 1988, Xalapa, Veracruz.

154. Francisco Urbina Soto, "Pluriprotesta," *Política,* October 4, 1988, 1.

155. Marco Antonio Aguirre, "Dicen las Madres Veracruzanas," *El Dictamen,* October 6, 1988, 5A.

156. Rosa Contreras Pérez, "Grupo antinuclear y ecologista: Temor de que la oposición use para su propio bien el lema 'No a Laguna Verde,'" *Diario de Xalapa,* October 10, 1988, 1.

157. Quoted in ibid.

158. Angeles Lizardi and Joaquín Rosas, "Resultados Electorales," *Política,* October 11, 1988, 1.

159. Ibid.

160. Irma Pilar Ortiz, "Inquietud entre los vecinos de Laguna Verde," *Excélsior,* October 11, 1988, 39A.

161. Interview with Marta Lilia Aguilar, December 5, 1989, Rancho Brazo Fuerte, Veracruz.

162. Ylia Ortiz Lizardi, "Antinucleares: Incertidumbre en Palma Sola," *Política,* October 12, 1988, 1.

163. Ibid.

164. Quoted in ibid.

165. Quoted in Daniel Ruíz, "Angustiosa espera de la decisión sobre Laguna Verde," *Diario de Xalapa,* October 12, 1988, 1.

166. Ibid.

167. Cristina Martín, "Alcudia ante comisiones senatoriales," *La Jornada,* October 12, 1988, 7.

168. Quoted in "Un día negro cuande se autorice," *El Dictamen,* October 12, 1988, 1.

169. Ibid.

170. Ibid.

171. Quoted in ibid. González Torres was referring to the presidential election in which the PRI candidate, Carlos Salinas de Gortari, defeated Cuauhtémoc Cárdenas. Many analysts question the accuracy of the election results.

172. Irma Pilar Ortiz, "Quedó ya lista para comenzar a funcionar," *Excélsior,* October 12, 1988, 1.

173. Rebeca Hernández Marín, "Reacciones en pro y contra del funcionamiento de la planta," *Uno Más Uno,* October 12, 1988, 1.

174. José Antonio Román, "Nada en concreto sobre la carga de la planta nuclear," *La Jornada,* October 12, 1988, 10.

175. The Chamber of Deputies—part of the legislative branch—is similar to the U.S. Congress in that it consists of two levels. Nevertheless, the executive branch is clearly dominant in the Mexican political system.

176. Miguel Angel Rivera and Emilio Lomas, "Piden dictámenes de seguridad," *La Jornada,* October 13, 1988, 1.

177. Ibid.

178. Fany Yépez, "Piden las Madres Veracruzanas consulta al pueblo antes de que funcione Laguna Verde," *Sol Veracruzano,* October 13, 1988, 1.

179. Ibid.

180. "Carta a los diputados y senadores," October 14, 1988, signed by twenty-two groups—including the Madres Veracruzanas, Grupo de los Cien, Pacto de Grupos Ecologistas, and Ganaderos Veracruzanos.

181. Ibid.

182. Ibid.

183. Ibid.

184. Alejandro Caballero, "Impugnaciones del FDN y el Grupo de los Cien," *La Jornada,* October 14, 1988, 15.

185. "Oficial: Laguna Verde en marcha," *Sol Veracruzano,* October 15, 1988, 1.

186. Quoted in Caballero, "Impugnaciones," 15.

187. Ibid.

188. Irma Pilar Ortiz, "Marchas y plantones alistan ya los grupos antinucleares," *Excélsior,* October 14, 1988, 1A.

189. Ibid.

190. Quoted in Francisco Mata, "El reactor está abierto y el combustible listo," *La Jornada,* October 14, 1988, 14.

191. Quoted in ibid.

192. "Contra la voluntad del pueblo: Laguna Verde se abre," *Diario de Xalapa,* October 15, 1988, 1.

193. A *jarocho* is a native of Veracruz. Quoted in Daniel Ruíz, "'Impida que funcione Laguna Verde y pase a la historia, Señor Gobernador!'" *Diario de Xalapa,* October 15, 1988, 1.

Chapter 4. The Loading of the Reactor

1. Carlos Medina and Rogelio Hernández, "Decide el gobierno poner en Marcha la nucleoeléctrica," *Excélsior,* October 15, 1988, 1A.

2. Ibid.

3. Ibid.

4. Editorial, "Laguna Verde, palo dado," *La Jornada,* October 15, 1988, 2.

5. "Los Cien: Grave, que hablen de abrir otras cuatro plantas," *Excélsior,* October 15, 1988, 1A.

6. Ibid.

7. Ibid. As stated earlier, the executive branch dominates the other branches of government in Mexico. Clearly, the movement organizations were making more of a symbolic stand here.

8. Madres Veracruzanas, joint declaration, quoted in ibid.

9. Interview with Rebeka Dyer, June 25, 1988, Xalapa, Veracruz.

10. "Editorial," *Política,* October 15, 1988, 1.

11. Ibid.

12. Ibid.

13. Carlos A. Medina, "La CFE: Ningún argumento sólido presentan los opositores," *Excélsior,* October 16, 1988, 1A.

14. Ibid.

15. Ibid.

16. Quoted in "Sobre Laguna Verde," *El Dictamen,* October 16, 1988, 1.

17. Madres Veracruzanas, *Diario de Xalapa,* October 16, 1988, Xalapa, Veracruz.

18. Quoted in Azucena Valderrábano, "Antinucleares le dan 'las gracias' al presidente," *La Jornada,* October 16, 1988, 1. Interestingly enough, the newspapers did not charge the antinuclear groups for the publication of these obituaries. Usually, paying for advertisements constituted a financial hardship for the groups.

19. Ibid.

20. "Ordenar la carga del reactor, acto autoritario, dice la oposición," *La Jornada,* October 16, 1988, 1.

21. Quoted in ibid.

22. Ibid.

23. Madres Veracruzanas of Emilio Carranza, quoted in ibid.

24. Ibid.

25. Quoted in Alfonso Valencia Ríos, "El gobernador intercambia opiniones con funcionarios de seguridad nuclear," *El Dictamen,* October 16, 1988, 1A.

26. Guillermo Zamora, "'¿Cuál plan de emergencia?' se preguntan lo Veracruzanos," *Proceso,* October 17, 1988.

27. Quoted in ibid.

28. Quoted in ibid.

29. Interview with the Madres Veracruzanas, October 18, 1988, Xalapa, Veracruz.

30. Ibid.

31. Ricardo Alemán, "Chocan vendedores ambulantes príistas y antinucleares," *La Jornada,* October 19, 1988, 1.

32. I personally witnessed this event.

33. Antonio Gil, "Gritos y golpes en la cámara de diputados," *Sol Veracruzano,* October 19, 1988, 1.

34. Enriqueta Cisneros, "Leyenda negra: Aquí hay gato encerrado!" *Sol Veracruzano,* October 19, 1988, 1.

35. Quoted in ibid.

36. Ibid.

37. "Los diputados dan por concluido el debate sobre la nucleoeléctrica," *Diario de Xalapa,* October 19, 1988, 1.

38. Ibid.

39. Gabriel Arellano López, "L.V., cosa juzgada?" *Gráfico de Xalapa,* October 19, 1988, 1.

40. Joaquín Rosas, "Cerrar calles abre el diálogo," *Política,* October 19, 1988, 1.

41. Ibid.

42. Daniel Ruíz, *Diario de Xalapa,* October 20, 1988, 1.

43. Quoted in Joaquín Rosas, "CFE: Ayer cargaron el reactor; el que tiene errores," *Política,* October 22, 1988, 1.

44. Ibid.

45. Ibid.

46. Emilio Lomas, "Anuncian la creación de cuatro plantas más con Laguna Verde," *La Jornada,* October 19, 1988, 1.

47. Angel Leodegario Gutiérrez, "Se recarga el reactor de la deuda," *Política,* October 19, 1988, 1.

48. Guadalupe H. Mar, "Soldados en Palma Sola: Antinucleares exigen que sean retirados," *Política,* October 25, 1988, 1.

49. Quoted in Joaquín Rosas, "Laguna Verde, no!" *Política,* October 24, 1988, 1.

50. Ibid.

51. Ibid.

52. Miguel Rico Diener, "El petróleo y Laguna Verde: Autoritarismo y neocolonialismo," *Uno Más Uno,* October 20, 1988, 3.

53. Interview, Madres Veracruzanas, October 20, 1988, Xalapa, Veracruz.

54. Carlos Jesús Rodríguez, "Opinión," *El Dictamen,* October 15, 1988, 1.

Chapter 5. Palma Sola

1. Interview with Marta Lilia Aguilar, December 5, 1989, Rancho Brazo Fuerte, Palma Sola, Veracruz.

2. Ibid.

3. Interview with Antonio Bretón, member of the Cattlemen's Association, July 21, 1988, Palma Sola, Veracruz.

4. Ibid.

5. Ibid.

6. Ibid.

7. Ibid.

8. Ibid.

9. Guadalupe H. Mar, "Soldados en Palma Sola: Antinucleares exigen que sean retirados," *Política,* October 25, 1988, 1.

10. Quoted in ibid.

11. Interview with Marta Lilia Aguilar, October 21, 1988, Rancho Brazo Fuerte, Palma Sola, Veracruz. I personally witnessed this incident. Military personnel also took a camera away from a journalist who was attempting to take photographs.

12. Interview with Marta Lilia Aguilar, December 5, 1989, Rancho Brazo Fuerte, Palma Sola, Veracruz.

13. Ibid.

14. Carlos Jesús Rodríguez, "Gutiérrez Barrios dialogó con los Ganaderos," *El Dictamen,* October 20, 1988, 1A.

15. Ibid.

16. Interview with Eduardo Gómez Téllez, June 26, 1995, Veracruz, Veracruz.

Chapter 6. The Antinuclear Movement Exposes Laguna Verde's Problems

1. Carlos Jesús Rodríguez, "Nada intimidó a los grupos antinucleares en la capital," *El Dictamen,* November 6, 1988, 1A.

2. Personal communication with the Madres Veracruzanas, November 22, 1988, Xalapa, Veracruz.

3. Ibid.

4. Ibid.

5. Angel L. Gutiérrez, "Hace simulacros sin avisar, CFE," *Política,* November 21, 1988, 1.

6. Editorial, "Cuando venga el lobo, qué?" *Política,* November 21, 1988, 1.

7. Ylia Ortiz Lizardi, "Los simulacros," *Política,* November 22, 1988, 1.

8. Daniel Ruíz, "Sí hubo accidente en Laguna Verde: Arias," *Diario de Xalapa,* November 23, 1988, 1.

9. "Los simulacros en L.V., deben mantenerse en secreto: El director de investigaciones nucleares sostiene que no se dé aviso a la población," *Diario de Xalapa,* November 5, 1988, 3.

10. "No ha sufrido ningún accidente: CFE," *Sol Veracruzano,* November 26, 1988, 1.

11. Guillermo Zamora, "Hubo un accidente en Laguna Verde el día 16; el simulacro fue después," *Proceso,* November 26, 1988.

12. Guillermo Zamora, "Negó el director datos básicos a los diputados investigadores," *Proceso,* December 5, 1988, 30–31. The Frente Democrático was a forerunner of the PRD. Not all members of the Frente joined the PRD when it was founded in 1989, however. See Kathleen Bruhn, *Taking on Goliath: The Emergence of a New Left Party and the Struggle for Democracy in Mexico* (University Park: Pennsylvania State Univ. Press, 1997), 324.

13. Zamora, "Negó."

14. Ibid.

15. Ibid.

16. Quoted in ibid.

17. Quoted in ibid.

18. Guillermo Zamora, "Otra grave falla; También la ocultó la CFE," *Proceso,* December 12, 1988, 30–31.

19. Quoted in ibid.

20. Ibid.

21. Ibid, 31.

22. Violeta Pacheco García, "Fuga radiactiva en Laguna Verde, denuncian Ganaderos," *Diario de Xalapa,* March 8, 1989, 1; Antonio Armenta, "Tres fugas radiactivas de L.V. en ocho días," *El Universal,* March 8, 1989, Sección en la Provincia, 3.

23. Ibid.

24. Martha Meza, "Encontradas opiniones sobre lo ocurrido en Laguna Verde," *Sol Veracruzano,* March 9, 1989, 1.

25. Ibid.

26. Quoted in Oscar Pedro Reyes, "Que suspendan Laguna Verde, claman las Madres Veracruzanas," *Diario de Xalapa,* March 9, 1989, 1.

27. Ibid.

28. Antonio Armenta Núñez, "Piden analizar el aire, por el escape de radiactividad en Laguna Verde," *El Universal,* March 9, 1989, Sección en la Provincia, 1.

29. Lourdes Galaz, "Citarán a funcionarios de seguridad nuclear para que declaren," *Excélsior,* March 11, 1989, 5A.

30. Ibid.

31. Quoted in Violeta Pacheco García, "Insisten los Ganaderos en una audiencia con C.S.G. para informarle sobre Laguna Verde," *Diario de Xalapa,* March 16, 1989, 1.

32. Gaudencio García Rivera, *El Universal,* March 25, 1989, Sección en la Provincia, 1.

33. Quoted in Violeta Pacheco García, "Sí hubo accidente en L.V., admite el director de CFE; sorpresiva visita realizó a la planta nucleoeléctrica," *Diario de Xalapa,* April 1, 1989, 1, 6.

34. Violeta Pacheco García, "Reconoce errores y fallas en L.V.: Fernández de la Garza," *Diario de Xalapa,* April 7, 1989, 1, 6.

35. Ibid.

36. "Laguna Verde causa inquietud," *El Dictamen,* April 7, 1989, 1.

37. Ibid.

38. Juan Rodríguez, "Demandan auditorías en la planta nucleoeléctrica de L.V.," *El Universal,* April 8, 1989, sección 2, 1.

39. Miriam Gracia, "Aumenta la energía eléctrica generada en Laguna Verde: CFE," *Sol Veracruzano,* April 28, 1989, 1.

40. "Detectan más irregularidades en la planta de L.V.," *Diario de Xalapa,* May 5, 1989, 1, 6.

41. Quoted in Carlos Jesús Rodríguez, "Laguna Verde no debe funcionar: López Meza," *El Dictamen,* April 18, 1989, 1.

42. "Jesús Aguilar: Laguna Verde ya causa estragos," *Política,* April 22, 1989, 1; Ricardo Blanco, "Derrames de crudo aniquilan a la fauna marina en costas de Veracruz," *El Nacional,* April 22, 1989, 12.

43. Carlos Medina, "No se suspendió el permiso de la CNSNS," *Excélsior,* May 17, 1989, 5A.

44. Quoted in ibid.

45. Ibid.

46. Claudia Gutiérrez de Vivanco, Leticia Toral Romero, and Sara González of the Madres Veracruzanas, quoted in Rosa Contreras, "Fugas radiactivas siguen alterando el ambiente de L.V.," *Diario de Xalapa,* June 5, 1989, 1, 6.

47. "Laguna Verde muestra de la ineptitud de la CFE, aseguran grupos ecologistas," *El Universal,* June 25, 1989, sección 1, 31.

48. Rosa Contreras Pérez, "Laguna Verde continúa sin funcionar por nuevas fallas," *Diario de Xalapa,* July 21, 1989, 1.

49. Ibid.

50. Rogelio Freyre, "Hay radiactividad a 500 kilómetros a la redonda de Laguna Verde: Madres Veracruzanas," *Excélsior,* August 1, 1989, Sección en los Estados, 2.

51. CFE paid announcement quoted in *La Jornada,* August 18, 1989.

52. Oscar Pedro Reyes, "Por dar información separan de su cargo a un alto funcionario de L.V.," *Diario de Xalapa,* August 22, 1989, 1; Alma Elena Gutiérrez, "Cesan a un funcionario de L.V. por proporcionar información," *Excélsior,* August 23, 1989, Sección en los Estados, 2.

53. Esaú Valencia Heredia, "Oculta la CFE fallas de Laguna Verde: Pescadores," *El Nacional,* August 23, 1989, 8.

54. Rogelio Freyre, "Oculta SEMIP la fuga radiactiva de L.V.," *Excélsior,* August 24, 1989, Sección de los Estados, 2.

55. Marco A. Martínez Negrete, editorial, "Foro de *Excélsior,*" *Excélsior,* April 10, 1989, 4A.

56. Oscar Pedro Reyes, "Millones de litros de agua radiactiva contaminan el mar," *Diario de Xalapa,* August 29, 1989, 1.

57. Eduardo Gómez Téllez, quoted in ibid.

58. Quoted in ibid.

59. Quoted in ibid.

60. Antonio Armenta, "Niega Laguna Verde polémica vigente," *El Universal,* September 6, 1989, Sección en la Provincia, 1.

61. Oscar Pedro Reyes, "Radiactividad en el puerto de Veracruz, admite la CFE," *Diario de Xalapa,* September 5, 1989, 1.

62. Leticia Rodríguez, "Laguna Verde contamina mariscos; Paran la pesca," *Sol Veracruzano,* September 7, 1989, 1.

63. Nidia Marín, "Sube la radiactividad ambiental en L.V.: Los Cien," *Excélsior,* September 8, 1989, 5A.

64. Ibid.

65. "Ninguna contaminación radiactiva en el golfo: La Armada," *El Nacional,* September 9, 1989, 8.

66. Espinoza, quoted in ibid.

67. Gaudencio García, "Niega Delgado que se haya arrojado al mar agua radiactiva de Laguna Verde," *El Universal,* September 17, 1989, Sección en la Provincia, 1.

68. Alfonso Valencia Ríos, "Pocos serios los datos de contaminación," *El Dictamen,* September 18, 1989, 1A.

69. Violeta Pacheco García, "La CFE se compromete a hacer cambios organizacionales en L.V.: Bello," *Diario de Xalapa,* September 20, 1989, 1.

70. "Relevante papel de la agrupación," *Sol Veracruzano,* September 22, 1989, 1; "Técnico nuclear comprobará probables fugas en la planta de L.V.," *Uno Más Uno,* September 22, 1989, 15.

71. Oscar P. Reyes, "Sospechan de sabotajes en L.V.," *Diario de Xalapa,* September 23, 1989, 1.

72. Sonia García, "Se sigue evadiendo la responsabilidad de informar sobre L.V.: Antinucleares," *Diario de Xalapa,* September 25, 1989, 1.

73. Ibid.

74. Roberto Lozada Rivera, "Requieren diputados información sobre las fallas," *Sol Veracruzano,* September 25, 1989, 1; Inocencio Valdés Vázquez, "Demandarán legisladores de Veracruz comparezca el director de Laguna Verde," *El Día,* September 25, 1989, 8.

75. Agustín Contreras Stein, "Cuidan más a Laguna Verde," *Gráfico de Xalapa,* September 27, 1989, 2.

76. Quoted in ibid.

Chapter 7. Movement Politics

1. Regina Martínez, "Crean la Coordinadora de 'No a Laguna Verde,'" *La Jornada,* February 20, 1989, 13.

2. The New Social Movements literature notes that groups in such movements

tend to be fluid and nonhierarchical. The Madres Veracruzanas did not join CON-CLAVE; they insisted on maintaining their autonomy from both the government and the other antinuclear groups—also a characteristic of the New Social Movements.

3. Raymundo Jiménez, "Concluyó reunión de grupos antinucleares," *Uno Más Uno*, February 20, 1989, 13.

4. Regina Martínez, "Crean la Coordinadora de 'No a Laguna Verde,'" *La Jornada*, February 20, 1989, 13.

5. Guadalupe H. Mar, "Martínez Negrete: Afectará las elecciones," *Política*, August 11, 1989, 1.

6. Ibid.

7. Interview with Rebeka Dyer, March 16, 1989, Xalapa, Veracruz.

8. Ibid. Dyer and her husband are now divorced.

9. Ibid.

10. Ibid.

11. Interview with Sara González, March 16, 1989, Xalapa, Veracruz.

12. Interview with Rebeka Dyer, March 16, 1989, Xalapa, Veracruz.

13. Ibid.

14. See Alberto Melucci, *Nomads of the Present*.

15. Oscar Pedro Reyes, "Las Madres Veracruzanas denunciarán ante C.S.G. la alta inseguridad de L.V.," *Diario de Xalapa*, November 14, 1989, 3.

16. Rosa Contreras Pérez, "Recibirá Salinas a las Madres Veracruzanas," *Diario de Xalapa*, November 19, 1989, 1.

17. Interview with the Madres Veracruzanas, November 14, 1989, Xalapa, Veracruz.

18. The account of the crisis is based on my direct observation of the Madres in November 1989.

19. Antonia Castillo Santos, "Las Madres Veracruzanas confían en C.S.G.," *El Dictamen*, November 26, 1989, 2A.

20. Regina Martinez, "L.V.: Primero, la auditoría," *Política*, February 26, 1990, 1.

21. Ibid.

22. "L.V.: Habrá auditoría," *Política*, March 1, 1990, 1.

23. Angel L. Gutiérrez, "Laguna Verde: Auditoría simple," *Política*, March 14, 1990, 1.

24. Quoted in ibid.

25. Ibid.

26. Regina Martínez, "L.V.: España auditará," *Política*, March 14, 1990.

27. Claudia Gutiérrez of the Madres Veracruzanas, quoted in ibid.

28. Ibid.

29. Alma Sámano Castillo, "Laguna Verde, la nucleoeléctrica más peligrosa del mundo," *Revelación,* September 10, 1990, 15.

30. Rodrigo Vera, "Sólo un simulacro de revisión, hecha por un amigo, se hizo para abrir la planta," *Proceso,* August 20, 1990, 7.

31. Ibid.

32. Ibid.

33. "Revisión técnica de la seguridad de la Unidad I de la central de Laguna Verde," document included in ibid.

34. Rodrigo Vera, "Sólo un simulacro de revisión, hecha por un amigo, se hizo para abrir la planta," *Proceso,* August 20, 1990, 7.

35. Quoted in ibid.

36. Ibid.

37. Quoted in ibid, 14.

38. Quoted in ibid, 18.

39. Quoted in ibid.

40. "Destina el gobierno federal más de 350 millones de dólares," *El Dictamen,* September 18, 1990, 1. That goal was later postponed to 1994.

41. Quoted in ibid.

42. Teresa Weiser, "Riesgos, apresuramiento en Laguna Verde," *Uno Más Uno,* December 8, 1988, 6.

43. Ibid.

44. Interview with Pedro Lizárraga, June 15, 1993, Xalapa, Veracruz.

45. Antonia Castillo Santos, "Plantón de los antinucleares," *El Dictamen,* April 10, 1989, 1.

46. Other mothers' groups in Latin America have not had this same protected experience, however. Several of the Madres de la Plaza de Mayo in Argentina disappeared, and others were tortured. See Marguerite Guzman Bouvard, *Revolutionizing Motherhood: The Mothers of the Plaza de Mayo* (Wilmington, Del.: Scholarly Resources, 1994).

47. Homero Aridjis, "Sobre Laguna Verde," *La Jornada,* February 11, 1989, 13.

48. Ibid.

49. Ibid.

50. Open letter signed by the members of 135 antinuclear groups, *La Jornada,* February 7, 1989.

51. Ibid.

Chapter 8. The Decline of the Movement

1. Rosa Contreras Pérez, "Entrevista de las Madres Veracruzanas con Alemán," *Diario de Xalapa,* 27 July 1991, 1.
2. Ibid.
3. Ibid.
4. Ibid.
5. Ibid.
6. Rubén Pabello Acosta, "Al margen de la noticia, carta abierta al Licenciado Alemán," *Diario de Xalapa,* July 31, 1991, 3.
7. "Niégase información sobre Laguna Verde, afirman Madres," *Diario de Xalapa,* November 27, 1992, 1.
8. Madres Veracruzanas, letter to the editor, *Política,* October 24, 1992, 2.
9. Ibid.
10. Guillermo Zamora, "Graves fallas en la planta nucleoélectrica de Laguna Verde," *Proceso,* January 25, 1993, 27.
11. Anonymous source, quoted in ibid.
12. Ibid.
13. Ibid.
14. "Puras mentiras atómicas, Madres Veracruzanas: Nada garantiza la información de CFE sobre Laguna Verde," *Política,* January 28, 1993, 1.
15. Quoted in ibid.
16. Ibid.
17. Quoted in ibid.
18. Quoted in "'*Proceso* no tiene bases,' P.C.H.: Son argumentos que no citan fuentes," *Política,* January 27, 1993, 1.
19. "No mata Laguna Verde, Luna Lastra: Ninguna posibilidad de accidente," *Política,* February 1, 1993, 1.
20. Quoted in ibid.
21. Rosa Contreras Pérez, "Habrá nueva supervisión a Laguna Verde," *Diario de Xalapa,* March 17, 1993, 1.
22. Mirna Benítez, quoted in ibid.
23. Quoted in Luis Ignacio Aparicio Romero, "Desde Alemania advierten del peligro de Laguna Verde," *Diario de Xalapa,* November 7, 1993, 1.
24. Ibid.
25. Quoted in Luis Ignacio Aparicio Romero, "Desde Bélgica apoyan también la lucha contra Laguna Verde," *Diario de Xalapa,* November 14, 1993, 1.
26. Interview with Mirna Benítez, March 23, 1996, Xalapa, Veracruz.
27. Pedro Lizárraga, "Las madres antinucleares en la contienda electoral," *Política,* June 1, 1994, 1.
28. Ibid.

29. Rosa Contreras Pérez, "Quieren audiencia con P.C.H. las Madres Veracruzanas, sí contamina Laguna Verde, reíteran," *Diario de Xalapa,* October 9, 1994, 1.

30. Quoted in ibid.

31. Quoted in ibid.

32. "Muchos ecologistas sólo llevaron agua a su molino y se olvidaron de nosotros," stated a cattlemen's organization member, quoted in Abel Hernández Santos, "Laguna Verde operará comercialmente en 1994," *Cuenta,* December 1993, 28.

33. Regina Martínez, "Compraron a los verdes," *Política,* February 12, 1993, 1.

34. Antonio Vivero Salas, quoted in ibid.

35. Quoted in ibid.

36. Ibid.

37. Melucci, *Nomads of the Present.*

Chapter 9. Mothers' Movements and Feminist Theory

1. See Linda M. Blum, "Mothers, Babies, and Breastfeeding in Late Capitalist America: The Shifting Contexts of Feminist Theory," *Feminist Studies* 19, no. 2 (summer 1993), 291–311.

2. Micaela di Leonardo, "Morals, Mothers, and Militarism: Antimilitarism and Feminist Theory," *Feminist Studies* 11, no. 3 (fall 1985), 599–617.

3. Ibid.

4. Ibid., 602.

5. Ibid., 612.

6. Ibid.

7. María del Carmen Feijoo, "The Challenge of Constructing Civilian Peace: Women and Democracy in Argentina," in Jane Jaquette, ed., *The Women's Movement in Latin America* (Winchester, Mass.: Unwyn Hyman, 1989), 64–72.

8. Ibid., 77.

9. Ibid., 78.

10. Ibid.

11. Ibid., 88.

12. Amy Swerdlow, *Women Strike for Peace: Traditional Motherhood and Radical Politics in the 1960s* (Chicago: Univ. of Chicago Press, 1993).

13. Ibid, 235.

14. Ibid.

15. Ibid.

16. Ibid, 239.

17. Ibid.

18. Ibid, 240.

19. Blum, "Mothers, Babies, and Breastfeeding," 300.

20. Sara Ruddick, *Maternal Thinking* (New York: Ballantine Books, 1989).

21. See chapter 7.

22. Interview with the Madres Veracruzanas, March 25, 1990, Xalapa, Veracruz.

23. For the pro-nuclear camp, the technology itself is gendered.

24. "I'm not taking over the [Governor's] Palace. I'll withdraw. We're here because of our families. We can't take such a risk. And we have to say this to the press. We have to avoid a situation in which the movement takes over the palace and everyone blames the Madres." Interview with Rebeka Dyer, October 21, 1988, Xalapa, Veracruz.

25. "Lower-class people fight hard; in a sense we're reaping what they've sown. And lower-class women fight hard; the men often put them out in front during demonstrations in order to discourage repression. But these women are in danger." Claudia Gutiérrez, during an interview with the Madres Veracruzanas, July 22, 1988, Xalapa, Veracruz.

26. Interview with Mirna Benítez, December 8, 1989, Xalapa, Veracruz.

27. Feijoo, "The Challenge of Constructing Civilian Peace," 78.

28. For excellent examples of the use of gender analysis in the study of social movements in Latin America, see Sonia Alvarez, *Engendering Democracy in Brazil: Women's Movements in Transition Politics* (Princeton: Princeton Univ. Press, 1990), as well as Vivienne Bennett, *The Politics of Water, Urban Protest, Gender, and Power in Monterrey, Mexico* (Pittsburgh: Univ. of Pittsburgh Press, 1995).

29. Interview with Mirna Benítez, December 8, 1989, Xalapa, Veracruz.

Chapter 10. Conclusion

1. According to Alan Scott, autonomy is "the insistence that the movement and those it represents be allowed to fight in their own corner without interference from other movements, and without subordinating their demands to other external priorities" (*Ideology and the New Social Movements* [London: Unwyn Hyman, 1990], 19). New Social Movements also keep their distance from the state according to Fernando Calderón, Alejandro Piscitelli, and José Luis Reyna ("Social Movements," 27).

2. According to Alberto Melucci (*Nomads of the Present,* 70), "In complex societies social movements develop only in limited areas and for limited periods of time. When movements mobilize they reveal the other, complementary face of the submerged networks. The hidden networks become visible whenever collective actors confront or come into conflict with a public policy. Thus, for example, it is

difficult to understand the massive peace mobilizations of recent years unless the vitality of the submerged networks of women, young people, ecologists and alternative cultures is taken into account. These networks make possible such mobilizations and from time to time render them *visible*."

3. "The result of this process of constructing an action system I call collective identity. Collective identity is an interactive and shared definition produced by several interacting individuals who are concerned with the orientations of their action as well as the field of opportunities and constraints in which their action takes place." Further, "In this sense collective action is never based solely on cost-benefit calculations and a collective identity is never entirely negotiable" (Melucci, *Nomads of the Present*, 34–35).

4. Joe Foweraker, *Popular Mobilization in Mexico: The Teachers' Movement, 1977–87* (New York: Cambridge Univ. Press, 1993), 179.

5. Interview, Marta Lilia Aguilar, December 5, 1989, Rancho Brazo Fuerte, Veracruz.

6. "Conclusion: Theoretical and Political Horizons of Change in Contemporary Latin American Social Movements," in Alvarez and Escobar, eds., *The Making of Social Movements in Latin America*, 318.

7. Chicago: Univ. of Chicago Press, 1982.

8. Ibid., 39.

9. Ibid., 40.

10. Ibid., 43.

11. Ibid., 45.

12. Ibid.

13. Ibid., 48.

14. Christian Joppke, *Mobilizing Against Nuclear Energy: A Comparison of Germany and the United States* (Berkeley, Calif.: Univ. of California Press, 1993), 11.

15. Ibid., 13.

16. Clearly, of course, I am not arguing that the contemporary German state is authoritarian.

17. See María Lorena Cook, *Organizing Dissent: Unions, the State, and the Democratic Teachers' Movement in Mexico* (University Park: Pennsylvania State University Press, 1996).

18. Ibid., 296.

19. Ibid., 301. See Alvarez, *Engendering Democracy*.

20. The case of the Madres de la Plaza de Mayo is a notable exception in that several of the mothers were among the disappeared.

21. Hellman, "Mexican Popular Movements," 126.

22. Ibid.

23. Peter H. Smith, "The 1988 Presidential Succession in Historical Perspective," in W. Cornelius, J. Gentleman, and P. H. Smith, eds., *Mexico's Alternative Political Futures* (San Diego: Center for U.S.-Mexican Studies, University of California, San Diego, 1990), 407.

24. Ibid.

25. Hellman, "Mexican Popular Movements," 127.

26. Cook, *Organizing Dissent,* 294.

Bibliography

Primary Sources

Cuenta, December 1993.
Diario del Istmo, February 20 to August 8, 1988.
Diario de Xalapa, March 1, 1988, to October 9, 1994.
El Día, July 28, 1988, to September 25, 1989.
El Dictamen, May 19, 1988, to September 18, 1990.
El Financiero, October 7 to October 17, 1988.
El Heraldo, October 12 to October 18, 1988.
El Nacional, October 15, 1988, to September 9, 1989.
El Sol de México, September 30, 1988, to July 22, 1989.
El Universal, June 25, 1988, to September 17, 1989.
Excélsior, February 29, 1988, to September 8, 1989.
Gráfico de Xalapa, June 9, 1988, to September 27, 1989.
La Jornada, June 3, 1988, to August 18, 1989.
Política, October 4, 1988, to June 1, 1994.
Proceso, October 17, 1988, to January 25, 1993.
Revelación, September 10, 1990.
Sol Veracruzano, May 17, 1988, to September 25, 1989.
Uno Más Uno, July 28, 1988, to September 22, 1989.

Secondary Sources

Alvarez, Sonia
   1990. *Engendering Democracy in Brazil: Women's Movements in Transition
          Politics*. Princeton: Princeton Univ. Press.
Alvarez, Sonia, and Arturo Escobar, eds.
   1992. *The Making of Social Movements in Latin America: Identity, Strategy,
          and Democracy*. Boulder, Colo.: Westview.
Bennett, Vivienne
   1992. "The Evolution of Urban Popular Movements in Mexico between 1968
          and 1988." In *The Making Social Movements in Latin America*, ed.
          Arturo Escobar and Sonia Alvarez. Boulder, Colo.: Westview.

1995. *The Politics of Water, Urban Protest, Gender, and Power in Monterrey, Mexico.* Pittsburgh: Univ. of Pittsburgh Press.

Blum, Linda M.

1993. "Mothers, Babies, and Breastfeeding in Late Capitalist America: The Shifting Contexts of Feminist Theory." *Feminist Studies* 19, no. 2 (summer): 291–311.

Bruhn, Kathleen

1997. *Taking on Goliath: The Emergence of a New Left Party and the Struggle for Democracy in Mexico.* University Park: Pennsylvania State Univ. Press.

Buerba, Carmen

1988. "De Pátzcuaro a Laguna Verde: La experiencia antinuclear en Santa Fé de la Laguna." In *Laguna Verde Nuclear? No Gracias,* ed. José Arias Chávez and Luis Barquera. Mexico City: Claves Latinoamericanas.

Calderón, Fernando, Alejandro Piscitelli, and José Luis Reyna

1992. "Social Movements: Actors, Theories, and Expectations." In *The Making of Social Movements in Latin America,* ed. Sonia Alvarez and Arturo Escobar. Boulder, Colo.: Westview.

Collier, Ruth Berins, and David Collier

1979. "Inducements versus Constraints: Disaggregating Corporatism." *American Political Science Review* 73: 967–86.

Cook, María Lorena

1996. *Organizing Dissent: Unions, the State, and the Democratic Teachers' Movement in Mexico.* University Park: Pennsylvania State Univ. Press.

Davis, Diane

1994. "Failed Democratic Reform in Mexico: From Social Movements to the State and Back Again." *Journal of Latin American Studies* 26: 375–408.

De Leon, David

1988. *Everything Is Changing: Contemporary U.S. Movements in Historical Perspective.* New York: Praeger.

Di Leonardo, Micaela

1985. "Morals, Mothers, and Militarism: Antimilitarism and Feminist Theory." *Feminist Studies* 11, no. 3 (fall): 599–617.

Escobar, Arturo

1990. "Imagining a Post-Development Era? Critical Thought, Development and Social Movements." Paper prepared for the international seminar CENDES, October 1–6, Universidad Central de Venezuela, Caracas.

Feijoo, María del Carmen

1989. "The Challenge of Constructing Civilian Peace: Women and Democracy in Argentina." In *The Women's Movement in Latin America,* ed. Jane Jaquette. Winchester, Mass.: Unwyn Hyman.

Foweraker, Joe

1990. "Introduction." In *Popular Movements and Political Change in Mexico,* ed. Joe Foweraker and Ann Craig. Boulder, Colo.: Lynne Rienner.

1993. *Popular Mobilization in Mexico: The Teachers' Movement, 1977–87.* New York: Cambridge Univ. Press.

Fox, Jonathan

1994. "The Difficult Transition from Clientelism to Citizenship: Lessons from Mexico." *World Politics* 46(2): 151–84.

García, María del Pilar

1991. *Estado, ambiente, y sociedad civil.* Caracas: USB/CENDES.

García Michel, Hugo

1988. *Más allá de Laguna Verde.* Mexico City: Editorial Posada.

Guzman Bouvard, Marguerite

1994. *Revolutionizing Motherhood: The Mothers of the Plaza de Mayo.* Wilmington, Del.: Scholarly Resources.

Haber, Paul

1996. "Identity and Political Process: Recent Trends in the Study of Latin American Social Movements." *Latin American Research Review* 31: 171–88.

Hellman, Judith Adler

1994. "Mexican Popular Movements, Clientelism, and the Process of Democratization." *Latin American Perspectives* 81(21.2): 124–42.

Joppke, Christian

1993. *Mobilizing Against Nuclear Energy: A Comparison of Germany and the United States.* Berkeley: Univ. of California Press.

Klandermans, Bert, and Sidney Tarrow

1988. "Mobilizing into Social Movements: Synthesizing European and American Approaches." In *International Social Movements Research,* ed. Hanspeter Kriesi, Sidney Tarrow, and Bert Klandersman. London: JAI Press.

McAdam, Doug

1982. *Political Process and the Development of Black Insurgency, 1930–70.* Chicago: Univ. of Chicago Press.

McCarthy, J. D., and M. N. Zald, eds.

1977. *Social Movements in an Organizational Society.* New Brunswick, N.J.: Transaction Books.

Melucci, Alberto

1989. *Nomads of the Present: Social Movements and Individual Needs in Contemporary Society,* ed. John Kean and Paul Mier. Philadelphia: Temple Univ. Press.

Mumme, Stephen P.

1992. "System Maintenance and Environmental Reform in Mexico: Salinas's Preemptive Strategy." *Latin American Perspectives* 72(19.1): 123–43.

Paya Porres, Victor

1994. *Laguna Verde: La violencia de la modernización.* Mexico City: Miguel Angel Porrúa.

Price, Jerome
    1982. *The Antinuclear Movement*. Boston: Twayne.

Ruddick, Sara
    1989. *Maternal Thinking*. New York: Ballantine Books.

Scott, Alan
    1990. *Ideology and the New Social Movements*. London: Unwyn Hyman.

Simonian, Lane
    1995. *Defending the Land of the Jaguar: A History of Conservation in Mexico.* Austin: University of Texas Press.

Slater, David, ed.
    1988. *New Social Movements and the State in Latin America*. Amsterdam: CEDLA.

Sloan, John
    1985. "The Mexican Variant of Corporatism." *Inter-American Economic Affairs* 38: 3–18.

Smith, Peter H.
    1977. "Does Mexico Have a Power Elite?" In *Authoritarianism in Mexico*, ed. Jose Luis Reyna and Richard S. Weinart. Philadelphia: Institute for the Study of Human Issues.
    1990. "The 1988 Presidential Succession in Historical Perspective." In *Mexico's Alternative Political Futures,* ed. W. Cornelius, J. Gentleman, and P. H. Smith. San Diego: Center for U.S.-Mexican Studies, University of California, San Diego.

Stepan, Alfred
    1978. *The State and Society: Peru in Comparative Perspective*. Princeton: Princeton Univ. Press.

Swerdlow, Amy
    1993. *Women Strike for Peace: Traditional Motherhood and Radical Politics in the 1960s*. Chicago: Univ. of Chicago Press.

Tourraine, Alain
    1983. *Antinuclear Protest: The Opposition to Nuclear Energy in France*. New York: Cambridge Univ. Press.

Index

About the Author

Velma García-Gorena is Associate Professor of Government at Smith College, where she teaches courses on Latin America and on the politics of race and class in the United States. She received her Ph.D. in political science from Yale University. Her current research interests include the women's movement in Mexico and grassroots organizing in the *colonias* of the Texas-Mexico border. She was born and raised in Mission, Texas.